Seventeen Simple Lectures
on
General Relativity Theory

Seventeen Simple Lectures
on
General Relativity Theory

H. A. BUCHDAHL

Australian National University

A Wiley-Interscience Publication

JOHN WILEY & SONS

New York • Chichester • Brisbane • Toronto • Singapore

Library of Congress Cataloging in Publication Data:

Buchdahl, H. A. (Hans Adolph), 1919-
   Seventeen simple lectures on general relativity theory.

   A Wiley-Interscience publication.
   Includes index.
   1.   General relativity (Physics)–Addresses, essays,
lectures.   I.   Title.

QC173.6.B8   530.1'1   81-11376
ISBN 0-471-09684-9   AACR2

Printed in the United States of America

10 9 8 7 6 5 4 3 2 1

PREFACE

Over the years, while seeking instruction from standard textbooks on basic
notions underlying the general theory of relativity, I have been confronted
with uncertainties and perplexities at every turn. These detailed lecture
notes are a by-product of a determined attempt to clarify, if not to resolve,
at least some of them. Although they can serve as a self-contained introduction to the subject, their alternative purpose is to function as a salutary
supplement to many of the conventional introductory expositions which are
available. In other words, though the raw novice may profit much from
them, they will perhaps be of greater benefit to those who have already
taken a first, or even a second or subsequent course in the subject. They are
naturally somewhat unorthodox in style, not only in their discursiveness,
but also on account of the intermingling of problematic aspects of the
particular theory in hand with those of physical theory in general. Questions
are raised in large numbers. Many of these remain unresolved, whereas the
answers to others are surely naive at times, for I am not a philosopher of
science. If doubts, uncertainties, and preconceptions are constantly displayed; if at first sight the usual introductory mathematical material seems
to be missing; if the words gravitation and curvature effectively make their
appearance for the first time in the last lecture; if attention is focused at
length on the role played by metalaws or regulative principles in the
establishment of the theory: all these and other unfashionable features are
matters of pedagogic intent. Still, much of the material usually to be found
in introductory accounts is also included, even if here, too, pedagogic
motivations have brought a number of modifications with them. For example, in the context of spatial spherical symmetry the vacuum equations are
solved so as to give directly what is in effect the Kruskal-Szekeres metric,
without an excursion via the Schwarzschild metric; a singular energy tensor
associated with this vacuum metric is explicitly exhibited; interior solutions
are exemplified by a metric representing a static gaseous sphere whose
equation of state makes physical sense, unlike that of the Schwarzschild

v

interior solution, which does not; the Kerr solution, instead of being merely quoted, is derived; and so on.

Inevitably the balance of the material included and the extent to which emphasis is placed on this topic rather than that may seem either harmonious or bizarre, depending on one's point of view. If they be regarded as bizarre I can only plead that these are lecture notes which bear a strong personal imprint of my style of teaching. They certainly are not intended to fulfill the purpose of a comprehensive textbook in which one would look for a detailed reference to every aspect of the theory and its empirical implications, to every formalism, to every mathematical device which might be relevant to it: their purpose is different, as a glance at the list of contents will confirm. By the same token, to describe them as "simple" seems to me to be appropriate, granted that this is not taken to imply a claim that they are therefore necessarily "easy." Of course, a specialist working on some branch of general relativity theory will find what I may say about it "easy," even superficial, and find any warnings not to take this or that for granted superfluous, for he or she will be well aware of the difficulties involved. Yet not everyone is a "specialist," and many casual conversations over the years have convinced me that I am not alone in being plagued by all manner of perplexities.

The beginning of the course may well be felt to be a greater hurdle to be overcome than the rest and therefore the first two lectures are shorter than the other fifteen which are of roughly equal lengths. They contain, I think, no technical material which has not previously appeared in the literature, but, since this is no textbook, I considered it inappropriate to give explicit references. None of the many quotations are ascribed to specific authors, for I may have misunderstood them. However, all the quotations in question are taken from one or another of the books listed at the end of the last lecture. Throughout I am concerned solely with the orthodox relativity theory: no alternative "theories of gravitation" are contemplated, nor, for good reasons, are implications of quantum mechanics taken into account.

As regards prerequisites, these are fairly modest. A general background of physics is of course taken for granted, but that is not to say that familiarity with all of the theories which are briefly mentioned is essential. If one has merely a cursory idea of what, say, the Born-Infeld electrodynamics or process thermodynamics is about, so much the better; if not, one can manage without this knowledge. On the other hand, on a more specific level, I assume, first, a fairly sound acquaintance with the special theory of relativity and its language and, second, a sound knowledge of Euclidean tensor calculus, characterized by the constancy of the components of the metric tensor which need be neither diagonal nor positive definite. As regards notation, it seemed best to set out relevant explanations in a separate appendix. This serves the dual purpose of providing a reminder of the terminology of Euclidean tensor calculus. No knowledge of non-Euclidean tensor calculus is assumed, the relevant mathematics being developed when it is required.

The idea that the preparation of an appropriate course of lectures might help to ameliorate some of my difficulties first occurred to me during a contemplative stay of a few months at the Weizmann Institute of Science, Rehovot, late in 1975. I take this belated opportunity of thanking the Institute and Professor Yigal Talmi in particular for the warm hospitality I received there. The opportunity to put the intention into effect finally came in 1979 while I was resident as a Fellow at Churchill College, Cambridge. The facilities of the Institute of Astronomy were made available to me by Dr. Martin Rees and for this I am most grateful to him. To Professor Hubert Goenner I owe a great debt, for despite other pressing commitments at the time˘ he read the original version of the manuscript. His many comments, whether critical or supportive, were of very great value to me. I took most of them into account as best I could, but I know that much remains of which he disapproves and for this I must take the entire blame. I also derived much benefit from various remarks made by several anonymous referees.

H. A. BUCHDAHL

Cambridge, November 1979
Canberra, April 1981

CONTENTS

other relations of the special theory listed. First remarks on the idea of distance. On the invariance of Maxwell's equations. Questions concerning the idea of invariance.

to non-Euclidean metric. The constant λ not excluded by $P1$. Further remarks on the energy tensor. Exact realistic metrics unknown. The incorporation of test objects. Model of a test particle. The geodesic equation recovered. The need for approximative schemes. One such scheme described generically. Linearized equations.

Seventeen Simple Lectures
on
General Relativity Theory

LECTURE 1

To begin these lectures I should surely first say something about their intended content and character. Straight away I am faced with a difficulty. Supposing I were to say something like this: "The objective of this course is to give a simple, introductory account of the general theory of relativity, a theory intended to form a framework for the description of physical systems including their gravitational interactions, in contrast with the special theory of relativity in which the latter is disregarded. The examination of underlying conceptions will take precedence over the presentation of self-contained mathematical structures or of extensive formal developments." Then although this account would reflect true intentions, no matter how inadequately, it is not really acceptable, for it uses a language which should not yet be used. There is explicit reference to "gravitation," yet this term is not to be introduced until much later—in the seventeenth lecture—when we shall be in a better position to understand its proper connotation, at any rate on the view taken here. Morover, granted familiarity with the special theory of relativity, so-called, the phrase "general theory of relativity" conveys no obvious meaning to the uninitiated. True, there is a manifest verbal implication that the "general" theory is some generalization of the special. Yet, in what sense is it a generalization, if it be one at all?

I may seem to be belaboring this question to some extent, but without further ado I choose it as an appropriate starting point. As soon as one tries to answer it one is confronted with new questions which demand an answer. Not only that, but one begins to realize just how much various contemporary attitudes which surround the general theory are influenced by all manner of preconceptions, be they semantic, epistemological, ontological or even ideological. By way of just one illustration, how might one react to the views expressed by one author who refuses with polemical vigor to call Einstein's later theory "the 'general theory of relativity' because the latter name is nonsensical" and who holds that Einstein did not properly understand his own theory because he failed to see "that in the new theory he had created the notion of relativity was not among the concepts subject to generalization"?

To begin with, a theory may be held to be a generalization of another theory under various circumstances but to characterize these is no simple matter. Consider some examples, each time contrasting a certain "special" theory with what would normally be agreed to be a generalization of it: Bohr's theory of the hydrogen atom and the theory of atomic spectra based, in part, on the Bohr-Sommerfeld quantum conditions; the geometrical optics of isotropic media and that of anisotropic media; equilibrium thermodyamics and process thermodynamics; Maxwell's theory and the electrodynamics of Born and Infeld. It suffices to take all these theories for granted here without, for the time being, subjecting the notion of a theory as such to closer scrutiny. At any rate, each of these pairs of theories, whatever the generic differences between them, satisfies two criteria: (1) there is a strong family resemblance between the conceptual frameworks of the two theories making up a given pair; and (2) under specifiable circumstances the "general" that is, generalized, theory "collapses" into the special theory, though possibly only in the sense of a limiting case. We might now be tempted to say of *any* pair of theories that one is a generalization of the other if and only if they jointly satisfy both criteria. The trouble with this proposal is that there is rather too much vagueness in the notions of "family resemblance" on the one hand and of "conceptual framework" on the other. Let us, however, not be sidetracked into trying to remedy these deficiencies now. We have at least two tentative criteria in hand for our use; and we fix firmly in our minds that *both* are to be satisfied. In particular, it is not enough for (2) alone to be satisfied. It would seem to me to be quite wrong to regard nonrelativistic quantum mechanics as a generalization of Newtonian particle mechanics, notwithstanding the existence of the correspondence limit. It would likewise be wrong to speak of statistical mechanics as a generalization of phenomenological thermodynamics, notwithstanding the existence of a thermodynamic limit. On the other hand the waters already begin to be slightly murky when one asks whether perhaps special relativistic particle mechanics might be regarded as a generalization of Newtonian particle mechanics.

With these rough and ready ideas in hand it is time to return to the question which gave rise to them in the first place. To answer it we first need to have an understanding of each of the two theories to which it refers. Why then are they known as the special and general theories of relativity? It is hardly good enough to say "because Einstein called them that." Why did he? What are they about? Why "special," why "general"? What is the connotation of "relativity" in each case, if "relativeness" is its dictionary meaning? Furthermore, if "relative" is taken to mean "not absolute" we are faced with various possible meanings which "absolute" might take. Finally, why do we here speak of theories? Perhaps, in some preferred sense of the term "theory," either the special or the general theory or both are not simply theories at all?

These, then, are some of the more specific question which confront us. To do full justice to them is, of course, not feasible within the compass of a few short lectures, if only because their ramifications are so extensive as inevitably to include problems which have been the subject of disputation for centuries. My best course of action would therefore seem to be this: after certain generic preliminaries to develop the basic outlines of the general theory while constantly bearing our unanswered questions in mind. To this end I shall neither pursue a historical approach nor feel constrained to follow contemporary fashions. With the formal framework of the theory established we can go on to consider a few simple special topics if only to round out the picture a little. Once this has been done the time will be at hand to review problems previously encountered: if not to solve them, then at least to take renewed note of them.

To start the ball rolling, let us decide whether we accept the special theory of relativity as a theory. To make any rational decision one has to have criteria on which to base it. In the case in hand we must evidently first agree as to the meaning we wish to attach to the term "theory." Of course, in doing so we should as far as possible conform to common usage. It would not do to understand by "theory" what most others would understand by "hypothesis," for example. On the other hand where there is no universal agreement how are we to know what is common usage? We have to be hardheaded and simply select one of the more commonly held positions relevant to physics. It goes something like this: a theory is an explanatory framework which consists of two parts. The first is an abstract logical calculus, a formal, for example mathematical, system. It consists, on the one hand, of sentences which involve primitive symbols and derived symbols defined in terms of these; and, on the other, the vocabulary and syntax of logic. Some of the sentences are taken as axioms, the rest are theorems deduced from them. This formal scheme defines the logical structure of the theory; but it has as yet no empirical content. This is supplied by the second part of the theory which is a set of rules, variously called "rules of interpretation," "coordinating definitions," and the like. They are essentially semantical rules which provide observational interpretations for at least some of the primitive and derived terms. The theory as a whole now has empirical content: axioms function as physical hypotheses, derived theorems as physical laws, and their validity is then testable by experiment.

Now all this is of course shamefully abbreviated—no mention is made of all sorts of provisos, deficiencies, ambiguities, and so on to which this schematic picture of a theory is subject. For example, to speak of "observational interpretation" is all very well, but just what is one to understand by this? In some particular case a symbol P which occurs in the abstract calculus might happen to be given the observational interpretation "pressure of a gas" and this might be held to be acceptable in as far as such a pressure is accessible to direct observation, say by reading a manometer. Should such a measurement, however, qualify as direct observation? Is there not some other theory involved, namely a theory of some measurement process with *its* abstract calculus and rules of interpretation? If so, should this "secondary theory" not have been absorbed in the "primary theory" in the first place? If not, what would the situation be if P happened to be interpreted as "proton"? Should we perhaps contemplate hierarchies of theories? There clearly are many perplexities here: I shall simply set them aside, at least for the time being. Still, one particular sin of omission needs to be rectified at once. Earlier I used the phrase "physical law" without comment. Now I must surely say something of what is to be understood by this.

Very briefly, I here take a law to be a universal statement which asserts the existence of a certain uniform connexion, that is, a statement that a physical phenomenon of a certain kind always occurs whenever certain conditions are met. It is worth putting this in a different way. A law, as here understood, alleges that, subject to relevant stipulations and conventions, whenever certain conditions are satisfied certain consequences will be found to obtain. It covers factual as well as possible cases, the distinction between these being exemplified by the notional law "planetary orbits are elliptical": not only is every observed planetary orbit elliptical but it must be the case that anything having certain specified properties—other than that of moving about the sun in an elliptical orbit—will in fact be found to move in this way. Analytic propositions, like those of pure mathematics, are not to be considered to be expressions of laws. In general the meaning of the term "law" is context-dependent. Perhaps with tongue in cheek, let me remind you that to the uninitiated a law is more likely to be a rule of behavior laid down by legislation. In that case he may well go on to inferences concerning the existence and nature of putative legislators. One must not scoff at this, as the history of science shows. At any rate, a physical law of the kind now contemplated is usually established by inductive generalization based on the scrutiny of sets of isolated facts relating to a given class of phenomena and taking note of patterns of regularity or uniformity which they may exhibit. Again all manner of difficulties appear, and these, too, I set aside. Nevertheless, I hope that we are now in a better position to reach a decision as to whether we should think of the special theory of relativity as a "theory."

Evidently we need to ask whether its interpreted sentences function as physical laws or not. Here we must not jump to conclusions. Consider the example of the theory of particle motion when no restriction on speeds is envisaged. Though this may be called a "relativistic theory" its laws are not part of *the* theory of relativity. The latter derives from the dual recognition that (1) no empirical significance can be attached to the phrase "uniform motion with respect to space" and (2) any measurement of the speed of light in vacuo will always yield the same result. All this we shall review in due course. At any rate, one may wish to replace (1) by a statement of the kind "any frame of reference moving with uniform velocity relative to an inertial frame is inertial and the form of the general laws governing physical phenomena is such as not to imply a generic distinction between different inertial frames." Likewise the statement (2) about the velocity of light might be replaced by a prescription concerning the measurement, or instrumental meaning, of spatial distance; and so on. This need not detain us now, for, as you already know, (1) and (2) or their equivalents may be jointly subsumed under the following proposition to which I shall return later: the differential equations governing the evolution of physical systems— of things and fields —must be invariant under Lorentz transformations. With the previous standard scheme of a theory in mind we have to look upon this as an interpreted sentence. It is, however, not a law, for the very idea of subjecting its validity *directly* to an experimental test does not make sense. The same is, I think, true of any other of its interpreted sentences to the extent that these will be concerned with the spacetime vocabulary required for the interpretation of specific theories. In short, one concludes that the special theory of relativity is *not* a theory. We might perhaps call it a "regulative principle" being concerned with the form of theories. In as far as it transcends the idea of a theory as we generally understand it, we may alternatively call it a metatheory. Of course, whenever convenient we shall nevertheless continue to speak of the "special theory of relativity," keeping the conclusions just reached at the backs of our minds.

It is worth pointing out that regulative principles of one kind or another are not particularly uncommon in physics. For instance, given the differential equations of some set of theories, one may adopt the regulative principle that the equations of each theory shall express the stationarity of some action integral, or else that they be of at most the second differential order. Again, one might for some reason impose the regulative principle that every field theory should be such as to admit the introduction of a well-defined local energy density of the field. Of course, the more such principles one introduces simultaneously—sometimes for no clearly discernible reason— the more one runs the risk of eventually finding some of them to be in mutual conflict. As an aside, one may even regard phenomenological thermodynamics as a metatheory *if* one interprets its primary function to be that of a selection principle for mechanical theories of heat—kinetic theory and classical and quantum statistical mechanics—in the following sense: for such a theory to be admissible it must exactly imply the differential relation $T dS = dU + dW$, where S is interpreted as metrical entropy, T as absolute temperature, and so on in the usual way, and furthermore the Entropy Principle and the Third Law must be accommodated.

Hitherto, having taken some familiarity with the special theory for granted I have allowed myself to refer to inertial frames and to spacetime without comment. In fact I shall need to return to these notions in some detail. They are full of perplexities many of which I shall not be able to resolve; and you will simply have to form your own judgments.

LECTURE 2

We might now begin to turn our attention directly to the general theory of relativity. Let us tread warily, however, for the picture is in certain respects rather confused. Right from the beginning popular opinion—a phrase not intended to have any pejorative overtones—seems to have viewed it as a recondite discipline, surrounded by an air of mystery. This irrational state of affairs is fostered, perhaps inadvertently, by those who maintain—and who has not read this?—that the theory of relativity has "introduced new conceptions of space and time." Conceptions of space and time belong to the realm of psychology but one would hardly wish to maintain that the theory of relativity introduces a novelty into psychology. That there are conceptual difficulties no one would deny. Yet those of quantum mechanics are as great if not greater, yet no popular clamor ever arose over that subject. Perhaps the situation just outlined can best be explained on sociological grounds, but we need not pursue this topic.

On a nonpopular level one immediately comes up against peculiar divergences of opinion which cannot be so easily disregarded. Already in the first lecture I quoted a view claiming that the name "general theory of relativity" is nonsensical and that in one important respect at least Einstein did not understand his own theory. Striking, too, are the contemporary attitudes toward what Einstein regarded as the twin pillars on which he erected the general theory, namely the principle of equivalence and the principle of covariance, both of which we shall have to talk about later at some length. For example, the view has been expressed that the principle of equivalence performed the essential office of a midwife at the birth of general relativity and that the midwife should now be buried with appropriate honors. As for the principle of covariance, it has often been said to be "physically empty," it being claimed that *any* theory can be made covariant under arbitrary transformations of coordinates. The twin pillars seem to have turned out to be mere feet of clay; but one is led to wonder how many hidden conventions

or presuppositions have played their part in generating this slightly be-
wildering conclusion. It would be easy enough to extend this recital of
conflicting stances which have evidently been adopted from time to time.
There is no need to do this as long as we take the implicit warning to heart.

When I earlier adumbrated a sketch of the idea of a theory I did not
claim that every theory *must* be presented in conformity with it. Indeed, it
may not be desirable to do so, for the task of forcing a given theoretical
structure into this straitjacket would be onerous indeed. Even were one to
succeed in this effort, one would have a form of presentation of the theory
which belongs to its paradigmatic stage. As a matter of fact, it is not
unusual to find the theory presented in a way which is reminiscent of the
standard sketch—commonly one speaks of an "axiomatic theory." Occasionally physicists also speak of the "construction of a mathematical model"
but this use of the term "model" is not easy to understand for it seems to
run counter to various meanings which are normally attached to this term in
physics. I think what is being meant is the construction of the first—formal,
uninterpreted—part of the theory as previously understood. However that
may be, axiomatic presentations pay scant attention to pedagogic interests
and are thus not well suited to introductory accounts. We therefore proceed
differently.

On a pretheoretic level phenomena present themselves to our conscious-
ness. We perceive things; within our visual field they may overlap but this
overlap is not usually permanent: change takes place. Inherent in this
knowledge of change is this: that we look now and look again later. That
this makes sense I must take for granted. In other words, I accept without
further question the existence of a "later than" relation in as far as of *any*
two occurrences e_1 and e_2 I can definitely say that "I saw e_1 later than e_2,"
or "I saw e_2 later than e_1," or "I saw e_1 and e_2 together", as the case may
be. Of course, the word "see" here must be taken with a grain of salt—
perhaps "register" or "be conscious of" would be more appropriate. The
relation is transitive, irreflexive, and asymmetric. With each of a sequence of
occurrences a real number t may now be consistently associated so that if e_1
and e_2 are any pair of the sequence the three alternatives just enumerated
correspond to t_1 being greater than, less than, or equal to t_2, respectively.
Evidently a high degree of arbitrariness is still inherent in this numbering—
one is strongly reminded of the construction of empirical entropy functions
in thermodynamics.

I have been rather guarded in attempting to talk only about perceptions
so far. However, neither am I inclined to maintain a position of pure
solipsism, nor do I want to run the risk of falling into psychological traps.
Let us therefore accept that when a given individual has a perception that he

might verbalize as "the mutual collision of two objects" then this is no mere personal state of mind, no hallucination, but there are indeed two objects which lead an existence apart from him and with which he is somehow able to communicate. Likewise other individuals will be able to communicate with the same two objects, and different individuals can mutually compare their experiences. Moreover, under appropriate circumstances an individual might be replaced by some registering apparatus. This is, for instance, done in part when an individual registers various collisions together with what he sees on the digital display of some electronic device. He finds, say, that later and later collisions are associated with larger and larger numbers. The latter may be taken to provide a particular labeling of the kind already described. Constant explicit reference to individuals is no longer required and it suffices to talk about the readings on the display of the electronic device. The latter is an example of what one understands by a clock. One says that it registers the "time" t not of the collision but of the receipt of some "message" emanating from it. One refers to the "later than" relation as a temporal relation and a clock provides a quantitative measure of the temporal apartness of the arrivals of distinct messages.

This sketch may seem to be rather heavy-handed. Still it lies in the nature of the problem that it is not easy to see what one is talking about when one is discussing time; and that is the point I want to emphasize. It does not seem possible to define time, for all proposed definitions inevitably suffer from circularity. They involve references to expressions like "before," "after," "moving," and "memory," all of which are already temporal expressions. Temporal notions are deeply embedded in the language we use, for instance in its tenseness. We cannot point at something and say "this is time." Again, one can at once reject phrases of the kind "time flows," for "flowing" already involves temporal ideas, quite apart from the fact that only substances flow. Evidently one has to be content with a linguistic investigation with the object of finding out how, according to what rules, temporal expressions are used. This is a major task and we are in no position to undertake it here.

It is wise at this stage to select from among the innumerable devices which might serve as clocks a particular realistic one, namely an atomic clock. This is a kind of mechanism in which the vibrations of a quartz crystal are controlled by a certain microwave transition of some nucleus such as Cs^{133}. This shall be our standard clock *by prescription*, and it is supposed to be reproducible as required. The advance of its register by a chosen number of units defines a standard time lapse, say one second; this therefore being a certain numerical multiple of the period of the chosen transition. It eventually turns out that unnecessary formal complications in the description of natural processes are avoided if one employs such a clock. Nevertheless, its use is circumscribed by certain conditions. Some of these are theory dependent and one seems to have no option but to pursue a sequence of closer and closer approximations, in the hope that these may have a limit. It is just under circumstances such as these that one may come to doubt the integrity of axiomatic theories.

Return now to the idea of the mutual collision of two objects already contemplated. If these, P and Q, say, collide at time t—as registered on a clock attached to P—then they were not in mutual contact at times earlier than t. One says that at times prior to t P and Q were "moving relatively to each other" or also "that the spatial relationship between P and Q was changing." Some comments are necessary here. First, the expression "P is moving" has here appeared as a relational term in the sense that by itself it is incomplete and must be supplemented by the answer to the question "relative to what is it moving?" This stands in contrast to a statement such as "P is liquid" which is complete and requires no supplementation. As for the paraphrase in terms of "spatial relationships," a new element enters here as soon as we imagine "spatial" to be the adjectival concomitant of the noun "space," for then we would seem to be required to say what the latter might mean.

An initial explanatory attempt proceeds by analogy with the transition from the qualitative, subjectively founded "temporal relationship" to instrumental "time." Previously we received messages merely passively, registering the times of arrival. Now, however, we send out messages—say isotropically emitted light pulses—from a given object P attached to a clock and arrange for these pulses to be reflected at other objects $Q_1, Q_2, \ldots$. The reflection of a pulse emitted from P at some initial time will be registered to arrive back at P at corresponding times $t'_1, t'_2, \ldots$, and to arrive from various directions. In fact we find not only that the time lapses $t'_1 - t, t'_2 - t, \ldots$ in general differ from each other but also that they and the directions depend on the choice of the initial time t. The whole routine may be repeated, taking light pulses to be emitted from any one of the objects $Q_1, Q_2, \ldots$ and registering times and time lapses on the clock attached to the object in question. The ancillary conditions which may have to be imposed to ensure that these procedures actually lead to results which exhibit some inner consistency need not concern us for the present. Suffice it to say that by pursuing a methodology of this kind we are trying to achieve our immediate aim of quantifying spatial relationships; so that we might then be able to talk about "spatial apartness" as a counterpart to the "temporal apartness" encountered earlier. Eventually we might thus be able conceptually to come to terms with "space" just as we did earlier with "time."

Unfortunately this won't do. We may, it is true, take $t'_1 - t$ under certain conditions as a measure of the spatial apartness of P and Q_1 at time t. These conditions are those to which I referred a short while ago: that $t'_1 - t$ shall be sufficiently small is just one important example. Granted all this, what should we say now? Our previous experience is of no help. We cannot say "$t'_1 - t$ is the space between P and Q_1 at time t," for instance, without running counter to the established usage of the various words which occur here. To refer to "space lapse" in place of "space" would be worse. We seem to be driven to saying something like this: "$t'_1 - t$ is the distance between P and Q_1 at time t." Apart from the fact that except under special circumstances no consistent significance can be attached to this statement the word "space" no longer appears. In any case, the additional information contained in our empirical knowledge of the directions in which light pulses leave and return has not been accommodated.

Perhaps, then, we should simply maintain that space is the set of all spatial relations, just as one may well hold "family" to mean the set of all familial relations such as "is the mother of" or "is the sister of." Here we have a possible pitfall in as far as one may deny this meaning of "family," taking this instead to be the set of individuals between whom certain, namely familial, relations exist. The pursuit of this analogy, however, soon leads to unacceptable conclusions. At any rate, even if we countenance the interpretation of "space" as the set of all spatial relations, we have to face various uncertainties even at the present level. First, what is one to understand by "all" spatial relations? Is this intended to refer merely to relations between actually existing objects, or does one wish to include spatial relations which would obtain between objects if these existed? Such a counterfactual notion is a matter of dispute and I leave it unresolved. Second, the interpretation at issue seems to run counter to the normal, everyday use of the word "space" as in phrases such as "moving through space," "an object in space," or "a lot of space." These seem to contain a metaphorical element with an implicit allusion to the notion of substance. So far we have, however, not encountered any grounds for thinking of space in substantival terms. After all, because there is the familial relation "motherhood" there is no reason to assume the existence of a peculiar substance, or entity of some kind, called motherhoodness. Alternatively, the everyday use of "space" may be taken to be a reflection of our intuition. We are hardly in a position to discount this, bearing in mind the attention we paid earlier to the perception of phenomena. Yet, were we to enter into any kind of examination of the idea of space in this context we would sooner or later be confronted with Kantian propositions of the kind "space is not an empirical conception which has been derived from external experiences" or "space is a necessary a priori idea which is presupposed in all external perceptions." To analyze these here is totally out of the question. We take consolation in the fact that they have in the past been accompanied by other claims of an a priori character which are no longer acceptable.

We shall have occasion later to return to some of the matters over which we have skimmed so lightly so that we may amplify them in their proper place. Now, however, the time is at hand for examining with some care a crucial notion, namely that of the *free particle*, mention of which I have so far avoided.

LECTURE 3

What are we to understand by a "free particle"? We have two questions here. First, granted that the particle is a material object, what are the circumstances under which a given material object is rightly to be called a "particle"? The answer is something like this: let it be agreed that the object is somehow open to sufficiently detailed inspection at all times of interest. Then it qualifies as a "particle" if it has unvarying structure and is sufficiently small. There is some deliberate vagueness inherent in both these conditions. The interpretation of the phrase "sufficiently small" is context-dependent but I assume that in any given situation it will be possible to assign a definite quantitative significance to it. Bear in mind that we are throughout operating on the macroscopic level. On a terrestrial scale, for example, we might specify an object not larger than a marble, or perhaps a grain of sand, to be "sufficiently small"; and one may translate any specification of this kind into the language of clocks and light pulses if this seems a desirable step to take. Evidently a limitation is being placed here on the *size* of an object, but it is to be taken for granted that this limitation is so devised that its mass—meaning its familiar inertial mass—is also "sufficiently small." As for the invariability of structure, this is mainly intended to convey that during the period of interest the object visibly maintains its integrity, that is to say, it does not divide into parts. In particular, it does not emit matter or radiation. If uncertainties remain, I find these more acceptable than an evasive definition in terms of "idealized structureless point objects." In any event, we shall have occasion to say more about the notion of a particle later on, especially in the eighth lecture.

The second question, namely, "when is a particle free?," remains to be answered. This can only be done by detailed stipulation. A particle P is free if the following conditions are satisfied. First, P is not in direct contact with any other material substances; in particular, no strings are attached, nor is it immersed in a gas or any sort of fluid. Second, no beams of particles or of electromagnetic radiation are incident upon P. Third, P is unaffected by electric or magnetic moments which any nearby substances may possess or by any electric currents in its neighborhood. Whether these conditions are satisfied in any given case is a matter of verification on a straightforward phenomenological level. In principle it requires merely direct inspection and tests involving switches and devices such as electrometers and compass needles. As a particular example, if changes in the motion of *P*—whatever this motion may happen to be—are found to be correlated with changes in the current flowing through a nearby coil, then *P* is not free.

We take it for granted that any uncertainties which remain here may be eliminated by subsequent refinement: a methodological standpoint which should by now be familiar. Of course, one may fall into error in supposing that the resolution of some particular uncertainty is merely a matter of "refinement." On the contrary, it may revolve about some crucial feature which, if dealt with carelessly, may give rise to contradiction or circularity. It is not, for instance, difficult to anticipate the following question. That a particle *P* is free presupposes, by stipulation, that it be not influenced by electric charges and so on residing on other bodies: why not simply stipulate that it be altogether unaffected by other bodies? The answer is this: such a stipulation is to be rejected since it is empty—it lies in the nature of things that it is in fact unrealizable. To illustrate this contention, contemplate a body *P* in the solar system. Casual inspection reveals that it is affected by the sun and all the other bodies. This is so once and for all and there is no way of essentially modifying this state of affairs. Whereas the absence of electromagnetic effects on *P* could be ensured, perhaps by neutralizing any electric charge residing on it, there is no corresponding device whereby the effects on *P* of other bodies of the solar system can be nullified, say by neutralizing some kind of "charge" or by a process of screening. This is a crucial piece of empirical evidence. In short, a particle is free under the conditions originally enumerated; and when it is not free we say qualitatively that it is subject to *forces*.

With the idea of the free particle established we can now say what is meant by a general *inertial frame of reference*. This is a set *O, A, B, C* of four free particles, each supplied with a standard clock, the set being in principle subject to the sole condition that a pulse of light leaving any one of the particles in a definite direction will subsequently encounter at most one of the others. The beginnings of the coherent description of spatial relationships or, to the extent that these are not unchanging, of spatiotemporal relationships, in quantitative terms are now at hand. In principle this might for example be done by means of operations with light pulses already described in the second lecture. In pursuing the history of a particle *P*, for instance, arrange a sequence of nondirectional light pulses to be emitted

from P, to be reflected at O, A, B, and C. The times t_O, t_A, t_B, t_C at which the reflections arrived back at P may be registered by the clock attached to P. These four numbers which belong to a particular pulse are then a particular set of *coordinates* of the emission of the pulse, or more briefly "of P." The particular *coordinate system* which is implicit here is exactly the correspondence established by the particular routine just described between the emission of light pulses from P and quadruples of real numbers.

The generality of these prescriptions is salutary, but it is most inconvenient for our immediate purposes and we therefore restrict it. First, we impose the overriding condition that the time lapse $\tau(:=t'-t)$ between the emission of a light pulse from O at time t and its arrival back at time t' after reflection at A shall be sufficiently small. Then so arrange matters that τ is independent of the time of emission t. Now, it certainly cannot be taken for granted that this arrangement can in fact be achieved; indeed, in general it cannot. It is a matter of empirical evidence, however, that it can always be achieved over a period τ^* which is sufficiently small though still large compared with τ. That this is so is a matter of crucial importance. How large τ^* might be in any particular case can only be decided by experiment. At any rate, it is to be clearly understood throughout that the following construction shall be contemplated only between times t and $t+\tau^*$, as registered by the clock attached to O.

The state of affairs which obtains as regards O and A when τ is independent of t is verbalized as follows: "O and A are fixed relatively to each other." Also, a mutual *distance* $\frac{1}{2}\tau$, measured in seconds, is here assigned to O and A. If at first sight we find it strange that whether or not O and A are relatively fixed should depend on the choice of a particular class of clocks, this is a consequence of not having freed ourselves of the notion of the identifiability of "points of space" with which, for the sake of argument, the ends of a measuring rod can be made to coincide. In fact one cannot identify a point of space in something like the sense in which one might single out a point of some substance; meaning now a sufficiently small constituent part of it. It is senseless to say "this point of space is now here and will later be there." Clearly, then, the question whether two particles are relatively fixed can only have a conventional answer and the adoption of different conventions will lead to divergent conclusions. If in accordance with our original stipulations O and A are relatively fixed they need not necessarily be so according to the alternative convention that they are relatively fixed if they can permanently coincide with certain marks engraved on a metallic rod. This is, so to speak, the customary convention of everyday life. There are pitfalls in this, however. For inherent in it is an assumption that the rod is in some sense "invariable." Yes, but in what sense? In the sense that its ends are relatively fixed, perhaps? No, that would plainly be a vicious circle. That it is not subject to changes of termperature? This has also been held to involve a vicious circle but that is, I think, a mistake. However that may be, let it be granted that any thermal or chemical or any other sort of clearly identifiable change has been eliminated. Then nevertheless the invariability of the rod is still a convention, for whether it does or does not obtain is not amenable to empirical test. On the other hand, the two alternative descriptions are in fact empirically equivalent. It is of course very convenient that this should be so, and had it turned out otherwise we might have been tempted to adopt another class of "standard clocks"; provided, however, that in so doing overcompensating complications would not result elsewhere as a consequence. Certainly we are already much less likely to be misled by whatever preconceived notions about "distance" we might have.

Generically, what has been done for A in relation to O is now to be done for B and C as well. The inertial frame thus restricted in character may be called a *local inertial frame*, but I shall often omit the qualification "local" unless confusion is likely to result. It will generally be useful to specialize it further, first, by so disposing its particles that the reflections by them of a light pulse from O return to O at the same time t' and do so in mutually perpendicular directions; and second, by synchronizing its four clocks according to the following prescription. Let a light pulse be emitted from O when its clock registers t. Then, when it reaches A, B, and C their clocks are to be set to register $t + \frac{1}{2}\tau$. The relation of synchrony so established is transitive, indeed it is an equivalence relation. In saying this I have taken into account the empirical result that A, B, and C, taken in pairs, are relatively fixed. As a matter of fact, we earlier assigned the mutual distance $\frac{1}{2}\tau$ to OA and to OB. By means of a light pulse exchanged to and fro between A and B in the way by now familiar we can empirically assign a mutual distance between A and B without reference to O. It turns out that, within the limits of experimental error, this distance is $\tau/\sqrt{2}$.

We need something more general than this. To this end, contemplate any particle P which is "at rest with respect to our inertial frame," meaning that it is fixed relative to its constituent particles. The particular way of attaching coordinates to P which I briefly described a little while ago was only intended to serve as an illustration. The first step toward a more convenient descriptive methodology is this: we experimentally directly determine— strictly speaking assign—the distances between P and each of the four particles O, A, B, C of the inertial frame. Let them be l, r_1, r_2, r_3 and write r_0 in place of $\frac{1}{2}\tau$. The second step consists of the replacement of r_1, r_2, r_3 by three numbers x^1, x^2, x^3 through the relation $(x^1 - r_0)^2 + (x^2)^2 + (x^3)^2 = r_1^2$

and the two others obtained from this by cyclic interchange of the labels
1, 2, 3. In fact, one so arrives at two alternative triplets of numbers. Experiment now shows that the relation $(x^1)^2 + (x^2)^2 + (x^3)^2 = l^2$ is satisfied by
just one of them: and we exclude the other. When the clock at P, supposed
synchronized with that of O, registers x^4 we assign to P the coordinates
x^1, x^2, x^3, x^4. A coordinate system defined by these prescriptions is called
Cartesian. The use of Cartesian coordinate systems, though by no means
mandatory, often leads to a desirable degree of formal simplicity. If, for
instance, two fixed particles have the coordinates x^1, x^2, x^3, x^4 and
$\bar{x}^1, \bar{x}^2, \bar{x}^3, \bar{x}^4$ then the mutual distance ascribed to them is $[(\bar{x}^1 - x^1)^2 + (\bar{x}^2$
$- x^2)^2 + (\bar{x}^3 - x^3)^2]^{\frac{1}{2}}$. When the $\bar{x}^a - x^a =: dx^a$ are sufficiently small—
"infinitesimal"—this mutual distance, dl, is given by

$$dl^2 = \delta_{ab} dx^a dx^b, \tag{3.1}$$

where $\delta_{ab} := \text{diag}(1, 1, 1)$. Had we adhered to the use of r_a in place of x^a a
more general quadratic form would have taken the place of the right hand
member of (3.1).

It is sometimes of advantage to pretend that the four basic particles of the
inertial frame are supplemented by a "sufficiently dense" cloud of free
particles all of which are mutually fixed. Each is attached to a clock and all
the clocks are synchronized with that of O. Any particular particle is thus
characterized by the values of its four coordinates. Any other particle Q
will, unless it is fixed, be in successive close proximity to a sequence of the
supplementary particles. With due allowance for the possibililily of interpo-
lation, the history of Q—its "motion"—is thus described by three functions
$\xi^a(x^4)(a = 1, 2, 3)$ in the sense that when Q happens to be at the supplementary particle P whose clock registers the particular time x^4 the other
three—"spatial"—coordinates are $x^a = \xi^a(x^4)$.

It is now a matter for experiment to determine the form of the functions
ξ^a when Q itself happens to be free. One finds that then, and only then, all
three are in fact linear. This is ultimately a reflection of a much simpler
situation involving just two particles both of which are free subsequent to
their mutual collision. If τ is the usual time lapse between the emission of a
pulse at time t from one of them and the arrival there after reflection of the
other, then observation shows that τ is a linear function of t. As regards any
given "light ray"—a pulse limited by suitable baffles—it will encounter a
sequence of supplementary particles. Once again, observation reveals that
the coordinates of these are linearly related: $x^a = \alpha^a w + \beta^a$, where w is a
parameter and the α^a and β^a are constants. Last, under the present
circumstances, a stretched string can be arranged to lie along the same set of
particles.

Especially with the relation (3.1) in mind, you will perhaps recall having read somewhere about "the validity of Euclidean geometry," a phrase often accompanied by claims to the effect that "space was thought to be Euclidean" but "that it eventually turned out to be non-Euclidean." Now, even though I am trying throughout to avoid emphasis on the language of geometry, the use of which so easily leads to terminological confusion and misleading analogies, a digression is required; for we have already agreed that one cannot identify "points of space" and this is enough to make the meanings of the phrases just quoted elusive, to say the least.

To begin with, it is wise to distinguish, at least temporarily, between two distinct meanings of the word "geometry," namely pure or abstract geometry on the one hand and applied or physical geometry on the other. First, what is pure geometry? It is a mathematical science, a purely abstract deductive system, consisting of a set of axioms together with a body of theorems derived from it; it therefore deals solely with implications. The set of axioms lays down relations which are to obtain between certain primitive terms, the only limitation being that the set must not be internally contradictory. They are essentially rules or definitions. They do not presuppose any intuition and to question their validity does not make sense, just as it would not make sense to question the validity of the analytic proposition "a bachelor is an unmarried marriageable male." To the extent that the theorems are deduced according to the rules of logical inference the geometry must therefore be valid as a matter of inner necessity. Any particular mathematical system one ends up with is fixed by the choice of a particular set of axioms. This choice is largely up to us. It follows that instead of asking "what is pure geometry?" we should have asked "what is a pure geometry?" For example, a certain set of 28 axioms generates *a* certain pure geometry, usually called plane Euclidean geometry. Now, however, we seem to be faced with another difficulty. Why do we refer to one but not to the other of a pair of mathematical systems as a geometry? I doubt whether a definitive answer to this question can be given at all. Presumably one speaks of a geometry because on traditional or analogical grounds it seems, by common consent, to be sensible to do so. Be that as it may, I repeat that no pure geometry has any empirical content. No concrete significance is to be attached to the occurrence of terms such as "points," or "straight lines". They are mere symbols—noises in the case of the spoken elaboration of a geometry.

Now, given an abstract geometry, one may seek to interpret its primitive terms, that is to say, correlate with them physical objects—not necessarily material—and the relations between them. As a result of this correlation the theorems of the pure geometry become statements involving physical objects. The whole interpreted scheme constitutes a "physical geometry" and whether or not this is valid is now a question of empirical test. Furthermore, we may legitimately speak of "a geometry" without the qualification "pure" or "applied," provided we then understand it from the start as a physical theory. According to our standard sketch the formal calculus which constitutes its first part is just a pure geometry. Of course, we still have the problem that we are not sure what class of theories should be called geometries, just as previously we could not say what mathematical systems should be call "pure geometries." Perhaps here, too, tradition plays its part.

Now, an attempt at a relevant, detailed interpretation of pure Euclidean geometry is difficult and I shall not attempt it. In what follows I shall sometimes use a language somewhat more customary than that used earlier in this lecture, but I assume that they are intertranslatable—it would be counterproductive to be enmeshed in superfluous pedantry. Still, on any level one soons gets into difficulties. For example, let me interpret a Euclidean straight line as follows: "a path is a straight line if and only if it is the path of a light ray." We might grant that this means something; but then what about the objection that there would appear to be no straight lines in the dark? Do we now have to talk about possible light rays? We are getting caught up in the puzzles surrounding modalities.

Nevertheless, let us adopt this interpretation, for we seem to be on relatively safe ground in the context of free particles. We can ascribe a length to the segment of a straight line in terms of the distance between two relatively fixed particles which are joined by a light ray which realizes it. We may go on to construct a right angled triangle by suitably disposing three mutually fixed particles—we already did so earlier. Then, as we saw, the "lengths of its sides" are found empirically to be subject to the Theorem of Pythagoras.

Are we now to say that "Euclidean geometry is valid"? Should we say that "space is Euclidean"? The answer to the first question is affirmative *if* the various interpretations of primitive terms are adhered to. This is an important qualification and I shall return to it in a moment, together with a reminder of an overriding condition imposed earlier. With regard to the second question we are in deeper waters. If one's ontological commitment is to space as the set of all spatial relations—we have encountered this before and will do so again—the answer to it must of necessity be the same as that to the first question. If one holds some other view of the nature of space one would seem to be faced with having to go from observations of the behavior of material entities to a theory about "space itself"; but it is not easy to see how this might be done.

The empirical knowledge of the "Euclidicity of space" is circumscribed by the particular interpretations which were adopted and the various stipulations contained in them. Now, granted these, suppose that the physical geometry had in fact turned out not to be Euclidean. In that case it would presumably be possible to adopt other interpretations and stipulations just so as to ensure the Euclidicity of the physical geometry. The latter then possesses the character of a mere convention. One may ask: why should one engage in such an endeavor? Is it just in order to satisfy prejudices conceived early in life? This hardly seems a sensible course of action, bearing in mind that it would inevitably inject immense complications into physical theory as a whole. However, instead of pursuing these matters here I shall return to them in the sixteenth lecture.

The reminder to which I referred a short while ago is this: the preceding considerations all related to *local* inertial frames in a sense previously defined; and if x^i are the Cartesian coordinates of any particle referred to such a frame, then the x^a must be sufficiently small. I shall refer to these limitations singly or jointly as the "condition of locality." Our conclusion as regards to the validity of physical Euclidean geometry is subject to its being satisfied; and one should therefore speak, more carefully, of its local validity.

LECTURE 4

In the course of the preceding lecture we made an effort to understand the conceptual basis of various conclusions which we reached. Among these were the following, always granted that the condition of locality is satisfied: the trajectory of a free particle is a straight line. It moves at constant speed, since its Cartesian coordinates x^a are linearly related to x^4; where by "speed" we here understand the quantity $[(dx^1/dx^4)^2 + (dx^2/dx^4)^2 + (dx^3/dx^4)^2]^{1/2}$. Any deviation from motion with constant velocity is then ascribed to the action of a force. Altogether, we have just Newton's First Law of Motion. In parenthesis, it would seem that we have not fallen into the trap of circularity and can ignore Eddington's paraphrase of the First law: "every body continues in a state of rest or of uniform motion in a straight line unless it doesn't."

We shall allow ourselves the luxury of regarding any material body to be made up of "elementary parts" which are to be looked upon as particles. The kinematic description of the motion of any particle amounts in the first instance to investigating the changes in the spatial relations between it and the local inertial frame of reference. On a dynamical level the changes are then to be correlated with the forces to which it is subject, the only phenomenological candidates being contact forces and electromagnetic forces. How this may be done is so well known that there is no need to go into any detail here. Let me, however, just issue the reminder that where a force enters into a law of motion which relates it to acceleration, then that force is not defined by the acceleration, that is, as rate of change of momentum, but must be prescribed extraneously, even if that prescription should involve the acceleration or its derivatives. At any rate, the description of other physical phenomena such as electromagnetic wave propagation is to be accomplished by reference to the same local inertial frame.

But wait! We must surely ask how we came to settle on this particular inertial frame $\mathcal{I}$? Is there anything special about it? On the face of it there is not, for the various procedural prescriptions were in the first place built around a set of arbitrarily chosen free particles O, A, B, C. There is no reason why they should not be applied to some other set of free particles O', A', B', C' in exactly the same way, giving us another inertial frame $\mathcal{I}'$. That the speed of light is always unity when light pulses are referred to $\mathcal{I}$ is built into the whole scheme from the outset. In this context there is, however, nothing which gives $\mathcal{I}$ preference over $\mathcal{I}'$. Therefore if light pulses are referred to the latter, the speed of light will of necessity again be unity. Of course, it would be rash now to conclude that $\mathcal{I}, \mathcal{I}'$ are intrinsically indistinguishable. Experiments had to be carried out previously the results of which eventually led us to assert the local validity of physical Euclidean geometry. We have, however, no automatic assurance that on the basis of results of observations referred to $\mathcal{I}'$ we could again maintain the local validity of physical Euclidean geometry. That we can in fact do so is an empirical result.

It would be a little dangerous to maintain that we should have anticipated this, perhaps arguing as follows: "after all, $\mathcal{I}$ was picked at random. For it to turn out to have a kind of privileged status would be too much of a coincidence." It is, however, a fact of life that one rarely proceeds at random—to some extent convenience often dictates what one actually does. It is just possible that considerations of convenience made us pick that frame which led to conclusions considered desirable; not forgetting, too, the possibility of ensuring Euclidicity by convention, a point already made toward the end of the last lecture.

Recall that both $\mathcal{I}$ and $\mathcal{I}'$ have to satisfy the condition of locality. In a convenient terminology, let us say that a particle P "is local with respect to a frame $\mathcal{I}*$" if the usual time lapse τ for the round trip of a light pulse from P to any particle of $\mathcal{I}*$ and back to P is sufficiently small. In particular $\mathcal{I}$ and $\mathcal{I}'$ will be local with respect to one another—or, simply "mutually local"—if O and O' are. As an intimation of things to come let me emphasize that hitherto we have nowhere required $\mathcal{I}$ and $\mathcal{I}'$ to be mutually local.

With this remark firmly implanted in our minds, let me then stipulate that until further notice any other inertial frame $\mathcal{I}'$ we may contemplate shall be local with respect to $\mathcal{I}$. This ensures that any free particle P local with respect to $\mathcal{I}$ travels in a straight line with constant speed, whether referred to $\mathcal{I}$ or to $\mathcal{I}'$, so that in this regard there is no generic distinction between the two frames. Likewise the description of the behavior of light pulses does not imply such a distinction. Physically light is, however, an electromagnetic quantity and we may therefore take it that the laws of electromagnetism, suitably formulated, will not imply a generic distinction between $\mathcal{I}$ and $\mathcal{I}'$. The mutual forces between charged particles express themselves through electromagnetic fields, the very notion of which derives from the existence of these forces. Self-consistency of the whole scheme would seem to require that the equations of motion of charged particles should also not allow a generic distinction to be drawn between $\mathcal{G}$ and $\mathcal{G}'$. The stage is set for an imaginative leap—a conjecture, it is true, but one with respectable antecedents. Subject to an empirical test of its validity, we make the following assertion: "the general laws governing physical phenomena are such that they do not accord a privileged status to any particular frame among the set of all mutually local inertial frames of reference." This is one form of the "special principle of relativity." Of course, the phenomena in question are themselves understood to satisfy the condition of locality, in a sense which should by now be clear.

I do not claim that the wording of the special principle of relativity just provided has no loose ends in it. It is notoriously difficult to state it in a foolproof way. Take the following statement which one occasionally encounters: "the laws of electrodynamics are valid in all frames of reference in which the equations of mechanics are valid." What can this mean? What if in one frame I have correctly written down the equations of mechanics but made a mistake in my presentation of the laws of electrodynamics? Even suppose I have the correct equations in both cases and I go over to some other frame by means of correct mathematical transformations: why should the resulting equations suddenly cease to be valid?

The situation is simply this. Suppose the differential equations governing the motion of particles on the one hand and those governing electromagnetic fields on the other, both referred to $\mathcal{G}$, have been written down correctly. Now make the transformations appropriate to the transition from $\mathcal{G}$ to $\mathcal{G}'$; that is, the fields and particles, whose objective existence we realistically took for granted from the start, are now referred to $\mathcal{G}'$. Then the resulting mechanical and electromagnetic equations must not contain the velocities of the particles of $\mathcal{G}'$ relative to those of $\mathcal{G}$, a demand which also governs other basic equations—think of Dirac's equation, for example. Sometimes one paraphrases all this by the requirement that the equations must be formally invariant under the transition from $\mathcal{G}$ and $\mathcal{G}'$; but either way one is perhaps verbally still on slightly slippery ground, a point to which I shall return shortly.

We seem to have reached a point beyond which we may proceed for a while with seven-league boots. This is because we are now reasonably prepared to take for granted a variety of consequences of the special principle of relativity as they are developed in a good many textbooks. The attitudes held by their authors and the starting points they adopt need be of no concern in the context of mere technical detail, for it is quite independent of them. Only the language used needs to be clarified and difficulties surrounding it noted. After that we can rest content with little more than a recital of specific results.

Earlier I had occasion to speak of "happenings" such as mutual collisions of pairs of particles. Of course, it has to be granted that there is some fuzziness even in this apparently simple idea. Particles are extended and have structure. One takes refuge in the notion of "idealization" and speaks of "point particles" and the like. In real terms, we simply ignore their extension and structure as long as we can. Put differently, we accept the idea of approximate spatiotemporal coincidence as useful. This we have done all along in order to be able at all to coordinatize a particle by only four numbers x^i. I shall call any physical phenomenon which can thus be satisfactorily coordinatized an "event." In particular, a "particle at a certain time"—with the appropriate meaning of time—becomes an event in this terminology. We have now landed ourselves in some difficulties as soon as we inquire into the meaning to be attached to the familiar term "spacetime." Sometimes this is taken to mean "the collection of all possible events," but that would mean that spacetime is somehow composed of all possible collisions, all possible particles, and so on. This is hardly acceptable if we adhere to our meaning of "event." Perhaps we can help ourselves by not taking the particles, collisions, and so on themselves, but rather their sets of coordinates as being events. That also is sometimes done, but events are now mere sets of numbers—a point to which I shall return shortly. Be that as it may, spacetime is occasionally taken to mean "the arena in which material processes are imagined to take place." This is very difficult to understand. We seem to have here a container picture of spacetime with substantival connotations. One's imagination is apparently called into play; we are implicitly expected to visualize the occurrence of physical processes by analogy with our everyday mental images of objects pursuing their histories. This won't do. The motion of a particle is an example of a material process. However, on either interpretation of an "event," the particle is a sequence of events in which its particular "motion" is already implicit. To say that this sequence "takes place in an arena" does not seem to make sense. The usual references to "particles moving along world lines" are equally objectionable.

Mindful of the brief examination of "space" undertaken in the second lecture, it is natural to contemplate "spacetime" as being the set of all possible spatiotemporal relations. A "point of spacetime"—an event, if you like—is then a subset of these relations. This is a bit loose; but, in any case, we are confronted again at least with all the difficulties and perplexities which we encountered before. This is not the place to rehearse them and we shall come back to them later.

I return briefly to the proposal, mentioned a moment ago, to take the set of coordinates of a "happening," rather than the happening itself, as being an event. On the face of it it has its merits since, after all, an event enters into the formal machinery of relativistic theory as a set of four numbers. Yet, are we not now in semantic difficulties? Is it right to speak of "a" set of four numbers here? Recall that a *coordinate system* is a correspondence between happenings and quadruples of real numbers; so that for every choice of coordinate system a given happening is, in general, characterized by a different set of four numbers. It seems we shall now have to say that an event is the set of all the coordinate quadruples of the happening which corresponds to all possible choices of coordinate systems. The prospect is not a happy one. Still, it will usually be clear what one has in mind when speaking of an event; it would be unduly pedantic to spell out the intended meaning on every occasion. It will be clear, for example, that the phrase "the event x^i" implies the prior choice of a definite coordinate system.

Evidently, what is now required is the relation between the sets of Cartesian coordinates x^i and $x^{i'}$ which characterize a given event E when referred to two arbitrarily chosen inertial frames $\mathcal{I}$ and $\mathcal{I}'$, these being mutually local, of course. It is worth emphasizing that while we are dealing with a transformation between coordinate systems—the quadruples x^i and $x^{i'}$ belong to alternative correspondences with E—this particular *coordinate* transformation is induced by the transition from one *reference* system to another. A coordinate transformation does not have to be—and in general will not be—of this special kind. At any rate, the solution of the problem in hand is too well known to need anything but the barest comment. One may arrive at it, for instance, by noting that (1) the front of a light pulse which is spherical when referred to $\mathcal{I}$ is also spherical when referred to $\mathcal{I}'$, (2) a free particle moves with constant speed in a straight line whether referred to $\mathcal{I}$ or $\mathcal{I}'$. The relations in question are the Lorentz transformations, that is, those transformations which transform the quadratic differential form

$$ds^2 := \eta_{ij}\, dx^i\, dx^j \left(\eta_{ij} = \mathrm{diag}(-1, -1, -1, 1) \right) \tag{4.1}$$

into itself. They are, in other words, the linear transformations

$$x^{i'} = L^{i'}{}_i x^i + L^{i'}, \tag{4.2}$$

where the $L^{i'}$ are four arbitrary constants, and the constants $L^{i'}{}_i$ have to satisfy the relations

$$\eta_{i'j'} L^{i'}{}_i L^{j'}{}_j = \eta_{ij}, \qquad \left(\eta_{i'j'} := \mathrm{diag}(-1, -1, -1, 1) \right) \tag{4.3}$$

leaving six of them independent. Altogether one has a ten-parameter group. The homogeneous Lorentz transformations can be constructed from successive simple Lorentz transformations of the kind

$$x^{a'}=x^a, \quad x^{b'}=x^b, \quad x^{c'}=\gamma(x^c-ux^4), \quad x^{4'}=\gamma(x^4-ux^c), \quad (4.4)$$

where $\gamma:=(1-u^2)^{-1/2}$, a, b, c is a permutation of $1,2,3$ and the constant u in general takes a different value each time.

As intended from the very beginning, I can now take for granted the terminology and indeed the formal machinery which derives from these results. As a reminder, the quadratic differential form on the right of (4.1) is the metric of spacetime, so called, referred to $\mathcal{I}$, and ds^2 is the interval between the events to which the coordinate differences dx^i are attached. For finitely separated events $\bar{E}, E$ the interval is $\Delta s^2 = \eta_{ij}(\bar{x}^i - x^i)(\bar{x}^j - x^j)$. A domain in which the metric takes the generic form (4.1) when a suitable coordinate system is used I shall again call Euclidean, though Lorentzian would perhaps be preferable. The interval is of course independent of the choice of inertial frame and so generates an absolute division of intervals into those which are timelike, spacelike, and lightlike according as Δs^2 is positive, negative, or zero respectively. From this derives the useful formal construction of the "light cone." The speed of material bodies or of any kind of signal cannot exceed unity—that of light—and one has the idea of causal connectibility. If the interval between E and $\bar{E}$ is spacelike there is no definite, that is, frame-independent, temporal order between them and the simultaneity of such events is a matter of prescription. There are various kinematic relations such as the so-called Lorentz contraction and time dilation, the composition of velocities, and the Doppler effect. One has the equivalence of mass and energy; and finally there are the appropriate formulations of the dynamical equations which govern the temporal behavior of systems of particles, of matter in bulk, and of fields such as the electromagnetic field, not forgetting the role played in these by energy-momentum tensors, by laws of conservation, and by principles of stationary action.

There is not much point in further extending this recital. Instead, I add some brief, incidental comments under just two of its headings which are likely to stand us in good stead later on. First, the Lorentz contraction. This is just an expression of the fact that in order to quantify spatial relationships at a given "time," reference to the particles constituting some inertial frame, say $\mathcal{I}$, is necessary; the time is then that registered by the synchronized clocks to which these are attached. The value of the mutual distance between two particles P and $\bar{P}$, fixed in $\mathcal{I}$, is an example of a specific measure attached to a spatial relationship. However, the previous prescription of -which it is the result requires suitable generalization if the same particles are to be referred to any other inertial frame $\mathscr{G}'$ such that O' and O are not mutually fixed. The value l' of the mutual distance assigned to P and $\bar{P}$ is now to be calculated by the previous rules from the differences between the coordinates $x^{a'}$ and $\bar{x}^{a'}$ of those particles of $\mathcal{Q}'$ with which P and $\bar{P}$ are approximately coincident at time $x^{4'}$—not x^4—that is, $l'^2 = \delta_{a'b'}(\bar{x}^{a'} - x^{a'})(\bar{x}^{b'} - x^{b'})$. The Lorentz transformation from $\mathscr{G}$ to $\mathscr{G}'$ then shows that l' is less than $l, l-l'$ depending on the relative velocity and orientation of $\mathscr{G}'$ and $\mathscr{G}$.

The point which I want to bring out here is that if the uninterpreted part of a theory—its grammar if you like—happens to contain the terms "distance" then its interpretation cannot be what inherited prejudices might incline one to regard as acceptable. A sentence of the kind "the distance between the ends of this rod is so many metres" is no more meaningful than the sentence "the velocity of this rod is such and such," for both must be supplemented by the phrase "with respect to such and such an inertial frame." I emphasize that this contention represents no a priori stance but is implicit in various empirical findings. It is just the utterance of an incomplete sentence of the kind "the length of this rod is so many metres" which implies a metaphysical acceptance of an entity called "length" which one can supposedly measure in various ways, each time with the same result. The world just isn't like that and it is wise always to bear that in mind. It would be futile to take the position that one might simply understand a length always to refer to a particular frame, that is, to contemplate so-called proper lengths alone, for that would be tantamount to declaring that spatial relations are unquantifiable except under very special circumstances. As for a naive operationalist position: we would apparently have to accept the conclusion that the length of a rod at rest in $\mathscr{G}$ and of the same rod not at rest in $\mathscr{G}$ are simply different physical quantities and that is not a happy prospect. More of this later in another context.

My remaining comments take Maxwell's vacuum equations as their starting point. In a coordinate system implicit in the form of (4.1) they are

$$f_{[ij,k]} = 0, \qquad \eta^{jk}f_{ij,k} = j_i, \qquad (4.5)$$

with the usual interpretation of the f_{ij} and of j_i in terms of the electromagnetic field vectors **E** and **B**, as measured with instruments fixed with respect to $\mathscr{G}$ and of the current and charge densities ascribed to charged moving matter. On referring these various quantities to an alternative inertial frame $\mathscr{G}'$, equations (4.5) go over exactly into themselves, that is, taking (4.3) into account, the new equations differ from the old solely by virtue of the primes attached to all indices. Here we must take careful note of the fact that $\eta^{j'k'}$ and η^{jk} stand for the same set of numbers. Were this not the case the explicit *form* of the equation connecting $f_{i'j'}$ and $j_{k'}$, referred to $\mathcal{I}'$, would be different from that of the equations connecting f_{ij} and j_k, referred to $\mathcal{I}$. The mere fact that the equations possess tensorial character—which they do with respect to arbitrary linear transformations—is not enough to ensure form invariance, granted that $\eta_{i'k'}$ is interpreted as the transform of the metric tensor. Form invariance will obtain only if $\eta^{i'k'}$ numerically coincides with η^{ik}, and (4.3) ensures that this will be so.

Strictly speaking we must adjoin to the equations (4.5) the equations of motion of a charged particle, that is, the relation between the Lorentz force and its rate of change of momentum. Only then will the back reaction of the field on its sources be known. This equation, too, and therefore the whole set of electromagnetic vacuum equations, is form invariant. It follows that if one investigates the field with measuring instruments fixed in $\mathcal{I}$ and again with instruments fixed in $\mathcal{I}'$ the quantities f_{ij} and $f_{i'j'}$ will be found to enter into formally identical equations. Therefore the $f_{a'4'}$ will be interpreted as the components of the electric intensity as measured with instruments fixed in $\mathcal{I}'$, and so on. Only now are we assured that the quantities $f_{i'j'}:=L^i_{i'}L^j_{j'}f_{ij}$, in the first place merely calculated by formal rules, do in fact have their customary significance.

The "transformed" electromagnetic equations make no reference to the orientation or velocity of $\mathcal{I}'$ relative to $\mathcal{I}$. One might object that the relations between $f_{i'j'}$ and f_{kl} do make such reference. This, however, is quite irrelevant, for the description of electromagnetic phenomena relative to $\mathcal{I}'$ is intended to be purely internal to $\mathcal{I}'$, that is to say, it is solely in terms of measuring instruments fixed in $\mathcal{I}'$. This remark disposes of a possible ambiguity which made us feel earlier in this lecture that we might be on slippery ground.

The second possible ambiguity concerns the meaning of "form invariance." This I already discussed a moment ago in the specific context of Maxwell's equations. It is, however, of quite some importance to examine rather closely what the phrase "form invariance of a differential equation"—or a set of such equations—is intended to mean. First, take again a special case, namely Laplace's equation for a scalar field U. If the coordinate system is Cartesian the equation is $\delta^{ab}U_{,ab}=0$, with $\delta^{ab}:=\text{diag}(1,1,1)$. This is manifestly invariant under orthogonal transformations and one's natural inclination is to consider it invariant only under these. Nevertheless Laplace's equation is often said to be "invariant under arbitrary transformations of coordinates." How can this be? To this question I shall turn my attention at the beginning of the next lecture.

LECTURE 5

To reduce the apparent terminological conflict noted at the end of the preceding lecture, recall first that when the coordinates are not Cartesian the metric is $g_{ab}(x^1, x^2, x^3)dx^a dx^b$, where the functions g_{ab} depend on the choice of coordinate system. Laplace's equation then has the generic form $F(U_{,ab}, U_{,c}, g_{de,f}, g_{gh}) = 0$ — the actual form of the function F need not concern us here. In any other coordinate system this becomes $F(U_{,a'b'}, U_{,c'}, g_{d'e',f'}, g_{g'h'}) = 0$ with the *same* function F. Now the g_{ab} and their derivatives are just certain functions of the coordinates, so that these two equations may be thought of as being written as $H(U_{,ab}, U_{,c}, x^d) = 0$ and $H'(U_{,a'b'}, U_{,c'}, x^{d'})$, respectively; but here the functions H and H' in general differ from each other. They will, however, be the same functions when the $g_{a'b'}$ are the same functions of the $x^{c'}$ as the g_{ab} are of the x^c.

Evidently what has happened is this. Initially Laplace's equation was held to be form invariant under a transformation of coordinates if and only if H' turned out to be the same function as H. By a semantic shift the equation was subsequently held always to be form invariant merely because the function F is fixed, that is to say, unaffected by any transformation of coordinates. Failure to distinguish between these alternative meanings of form invariance has led to endless confusion in the literature, especially as the terms invariance or covariance are also used in this context with vague nontechnical connotations. Therefore, when necessary I shall avoid ambiguity by speaking of proper and improper form invariance, respectively.

The issue is sufficiently important to warrant another explicit example which will also serve to illustrate an additional point. Take the eikonal equation $W := g^{ij}S_{,i}S_{,j} = 0$, governing some scalar field S. The transformed equation is $W' := g^{i'j'}S_{,i'}S_{,j'} = 0$. Manifestly W' is the same function of $g^{i'j'}$ and $S_{,i'}$ as W was of g^{ij} and $S_{,i}$: the equation is improperly form invariant. As regards proper form invariance on the other hand, care must be exercised here. The *expression* W is properly form invariant if and only if the $g^{i'j'}$ are the same functions of $x^{k'}$ as the g^{ij} are of the x^k. For the *equation* to be properly form invariant it suffices, however, for the $g^{i'j'}$ to be as just stated merely to within a common nonvanishing factor, since such a factor can simply be removed from the equation.

It is hardly necessary to go on to a detailed formal statement of the distinction between proper and improper invariance for the case of more general equations or sets of equations. Suffice it to say that the presence in them of auxiliary functions like the g^{ab} is of course necessary if improper invariance is not to be an empty notion. Such quantities are simply prescribed in one coordinate system and then the corresponding quantities in any other coordinate system are calculated according to specified rules. For example, $g^{i'j'}=L^{i'}{}_{i}L^{j'}{}_{j}g^{ij}$, where $L^{i'}{}_{i}:=\partial x^{i'}/\partial x^{i}$, since the g^{ij} are the components of a symmetric contravariant tensor—strictly speaking tensor field. Now suppose one wanted to write down a set of improperly form invariant second order linear partial differential equations for a field of some kind. This might, for the sake of argument, turn out to be impossible, even given the g_{ij}. Then, is there anything to prevent us from introducing further auxiliary fields with appropriately devised rules of transformation just so as to ensure the improper form invariance in that? No, there isn't. In fact, all tensor equations—but not they alone—are automatically improperly form invariant. One often comes across the requirement that certain field equations be "covariant"—presumably meaning improperly form invariant?—and this demand is then rightly held to be of no apparent "physical significance," on the grounds that one can always satisfy it, for example by constructing tensorial equations with the aid of a sufficient number of auxiliary tensors of one type or another. However, we are inevitably faced with having to conjure up new quantities which either lead an independent existence, that is to say, to which in some chosen coordinate system values are assigned by fiat, or else they will have to obey field equations of their own. Either way these new theoretical elements should have their observational counterparts. What if we do not find them? Then we shall of necessity be reduced to the requirement of proper form invariance; and this may turn out to be a very severe demand indeed.

What then of the sometimes prominently displayed claim that covariance of equations in itself is not the expression of a physical law? Presumably what is being asserted here is that the demand for covariance does not constitute a law, for covariance, when it obtains, is merely a fact. In as far as it is a condition imposed on the form of equations it is not a law in the first place, but a law about laws, that is, a metalaw. Does this, then, have no physical consequences? If covariance here means proper form invariance once cannot assert in generality that it does not; for, in the presence of other conditions which may have to be satisfied, it may in fact determine the form of the field equations of some theory uniquely. These in turn entail conclusions the validity of which is subject to observational tests. Why then should one hold the demand for "covariance"—properly understood—to be physically vacuous? I think one should not. At any rate, we shall have occasion to refer back to these ideas in a specific context.

Before moving farther afield let us return briefly to the phrase "special theory of relativity." About this I asked several questions already in the first lecture. These can be paraphrased in a nutshell as follows: why theory, why special, why relativity? As regards the first, I have already argued that the term "metatheory" is to be preferred here, since it is about formal aspects of theories. There is no need to rehearse that argument again. Next, why special? Perhaps to distinguish the special from the "general" theory? That may be true, but this facile answer is likely to land us in circularity when we come to carry through the intended examination of the relation between the two. I therefore take the view that the qualification "special" draws attention to the inclusion of the underlying premise that any local inertial frame is indefinitely extendible and should be taken to have been globally extended. Finally, "relativity" emphasized that the description of motions necessarily requires material referents and that, if these are chosen to be inertial frames, then none enjoys a privileged status among them. In particular, field equations must be properly form invariant under transitions from one such frame to another.

You will have noticed that I have studiously avoided the use of the term "absolute." This is because even in a restricted context it may have one of a variety of accepted meanings. Perhaps the four most relevant here are (I) the metaphysical meaning: capable of being thought of or conceived by itself alone, or existing without relation to any other being; (II) unchanging, fixed, and if not absolute in this sense, dynamical, that is, subject to evolutionary laws; (III) not relational, that is, not characterized by or taking its meaning from relations between entities; (IV) not relative, relative here meaning determined only by relation to something else or to some external standard. When advisable these four meanings, taken in order, may be distinguished from each other by attaching a superscript I,...,IV, as the case may be. Two examples are the metric $ds^2 = \eta_{ij}dx^i dx^j$ which is absolute[II] and the absolute[IV] zero of temperature. One has to treat absolute[II] with some circumspection, but we will worry about that when the time comes.

I have hitherto laid great emphasis on the so-called First Law of Motion. It seems appropriate to take this as the starting point for further progress. To begin with, note that it is a kinematical law, that is to say, it does not ascribe a cause to the constant motion of a free particle in a straight line. That this kind of motion obtains must simply be taken as a brute fact. Of course, had we chosen different stipulations and conventions on the way we might have ended up with motion of a different character; then this would have to have been accepted as a brute fact. Either way we are concerned with a "standard motion" and it seems to me that to search for a cause of this will inevitably result in vicious circularity. Be that as it may, one needs to know the significance of a standard motion, for only then can one speak meaningfully of nonstandard motion and ascribe deviations from the standard to causes—or forces—be they electromagnetic, nuclear, contact forces, or the like: but that is where we came in. At any rate, the laws of motion which we are talking about have purely local character, for their relevance is restricted to domains circumscribed by maintenance of the condition of locality.

Now, of necessity this condition as a whole cannot always be satisfied. For example, if we wish to describe the history of some particle P we may initially refer it to some local inertial frame $\mathcal{G}_1$. Then it may certainly happen that even during a period much less that τ^*, P will cease to be local with respect to $\mathcal{G}_1$ and, in any event, $\mathcal{G}_1$ itself will eventually cease to satisfy the condition of locality. In other words, we are in this way able to describe only a small part of the history of P. Certainly we can subsequently refer P to a new local inertial frame $\mathcal{G}_2$; but again we can cover only a small part of its history. Moreover, these fragmentary histories are disjoint, for $\mathcal{G}_1$ and $\mathcal{G}_2$ were selected quite separately and we know of no relation between them. It seems that we have no option but to go back, even if only provisionally, to a scheme of the kind already proposed in the second lecture when we were first investigating how one might quantify spatial relationships. To be specific, let $Q_i(:=1,\ldots,4)$ be any four particles which may, but need not, be inertial. At any event E of P let a light pulse be emitted. Its reflexion by Q_i reaches P after time lapses τ_i. These four numbers can be used uniquely to label E. There may be difficulties with degenerate cases, but these need not detain us. At any rate, we may take the τ_i as coordinates of E. Since the history of P is a continuous sequence of events we thus have in principle a means of describing—or coordinatizing—this history.

Whatever the conceptual merits of this proposal might be it is totally unrealistic, if only because it is difficult to see how one might arrive at any definite lawlike conclusions in a situation of such enormous complexity. I do not intend to spend any time on it, since it is in any case largely irrelevant. Its purpose here is to lend an air of respectability to what follows. First, the conclusion that a correspondence can be established between the events which constitute the history of any particle and a continuous sequence of quadruples of real numbers is generalized by the assumption that a correspondence can be established between the events in some, in general nonlocal, neighborhood of any given event and quadruples of real numbers. From the fourth lecture we recall that such a correspondence is just a coordinate system. Bearing in mind that we may think of any event as being part of the history of an unlimited number of particle histories, one may well feel justified in taking certain topological preconditions to be satisfied; and likewise one is inclined to assume that various functions which one encounters are sufficiently often differentiable. We could now begin to talk about "spacetime as a differentiable manifold" or as a "Riemannian manifold" and so on. I shall not do so for it would not enlighten us at this stage.

What seems more useful is to point out emphatically that I am at the moment not concerned at all with a manner of selection of particular coordinate systems or the interpretation of any such system in terms of measurement. The crucial point is that a coordinatization, reflecting continuity and so on, should in principle be possible at all. One frequently gains the impression that there is a demand for the direct interpretability of coordinates. It is not at all obvious why this should be so. In quantum mechanics few people worry about the fact that the wave function does not receive an immediate interpretation: the physicist knows what to do with it, how it enters into his calculations, as the result of which he makes testable predictions. To repeat, I rigidly maintain the standpoint that a coordinate system is no more than a correspondence between events and quadruples of real numbers. All that is taken for granted is that in principle such correspondences can be established, with topological and continuity conditions playing their part in governing their generic properties. The existence of a coordinate system certainly does not imply any metrical properties of spacetime. What, then, is one to make of the typical assertion that "a spacetime may have more structure than the coordinate system used to describe it," which may be found in the literature?

Be that as it may, whatever coordinate system x^i one happens to have in hand, there is no way so far of giving it preference over any other coordinate system obtainable from it by an arbitrary nonsingular sufficiently often differentiable transformation $x^i = f^i(x^{j'})$. There is no criterion at hand on which a choice of a special coordinate system might be based—no explicitly known functions appear which, perhaps for purely practical reasons, one might wish to reduce to a particularly simple form.

All along we have taken for granted the existence of spatiotemporal relations to which lawlike expression may be given. We have seen in some detail what this involves on the local level, the endpoint of the investigation being the "metrical relation" represented by the invariant metric $ds^2 = \eta_{ij} d\bar{x}^i d\bar{x}^j$. Local coordinates $\bar{x}^k$ have been suitably chosen and the origin of coordinates can be taken anywhere in a small but finite region. How are we now to proceed on the nonlocal level? I use the term nonlocal in preference to "global" since the latter would have unintended connotations here. On this level we have so far merely endowed spacetime with a coordinate system but still treated it as metrically amorphous. We have just reminded ourselves that it is in fact not so, for two neighboring events whose coordinates differ by dx^i when referred to a nonlocal system are separated by the invariant interval whose square is $\eta_{ij}d\bar{x}^id\bar{x}^j$. Bearing in mind that this state of affair obtains for any such pair of events, one is led to the following compelling assumption: the spatiotemporal relations between events are characterized by a nonsingular symmetric covariant tensor field $g_{ij}(x^k)$—the "metric tensor"—in the sense that the interval ds between arbitrarily selected neighboring events x^k and x^k+dx^k is given invariantly by $ds^2 = g_{ij}dx^idx^j$. If this is to be consistent with the local description of a neighborhood of a given event $\overset{0i}{x}$ it must always be possible to find a transformation to coordiantes $\bar{x}^i$ such that $\bar{g}_{ij}(\bar{x}^k)=\eta_{ij}+$ terms at least quadratic in $\bar{x}^k$; the absence of linear terms on the right ensuring that Euclidicity will obtain over a small but finite region. In future I shall call such coordinates, with origin $\bar{x}^k=0$ at P, normal coordinates at P. Their most important feature is that $\partial\bar{g}_{ij}/\partial\bar{x}^k=0$ at the origin.

I shall now show that such coordinates can always be introduced, pathological cases aside; but these we have implicitly disregarded anyway since they correspond to the impossibility of constructing a local inertial frame in the first place. Consider transformations of the form

$$x^i=\overset{0i}{x}+a^i{}_j\bar{x}^j+\tfrac{1}{2}b^i{}_{jk}\bar{x}^j\bar{x}^k+O_3, \qquad (5.1)$$

O_n standing for terms at least of degree n in the $\bar{x}^l$. The $a^i{}_j$ and $b^i{}_{jk}$, symmetric in j and k, are constants to be determined. The transform $\bar{g}_{ij}(\bar{x}^l)$ of $g_{ij}(x^l)$ is given by

$$\bar{g}_{ij}(\bar{x}^l)=(\partial x^m/\partial\bar{x}^i)(\partial x^n/\partial\bar{x}^j)g_{mn}(\overset{0l}{x}+a^l{}_p\bar{x}^p+O_2).$$

The right hand member of this may be explicitly evaluated by a straightforward calculation. Its constant term shows that $a^i{}_j$ must be so chosen that

$$a^m{}_ia^n{}_jg_{mn}(\overset{0k}{x})=\eta_{ij}. \qquad (5.2)$$

The terms linear in $\bar{x}^k$ will be absent provided

$$b^k{}_{ij}=-a^m{}_ia^n{}_j\overset{0}{\Gamma}{}^k_{mn}.$$

On the right $\overset{0}{\Gamma}{}^k_{mn} := \Gamma^k_{mn}(\overset{0}{x}{}^l)$, where

$$\Gamma^k_{mn} := \tfrac{1}{2} g^{kl}(g_{lm,n} + g_{ln,m} - g_{mn,l}) \tag{5.3}$$

is a "Christoffel symbol." Since g_{mn} is symmetric, the a^i_j can in fact always be so determined as to diagonalize it in the sense of (5.2), but they are, of course, fixed only to within an arbitrary Lorentz transformation. The signature of g_{ij} must evidently always be -2.

It is instructive to go a little further. Having removed the terms linear in $\bar{x}^k$ from $\bar{g}_{ij}$, we might try to remove the quadratic terms as well. To this end replace O_3 in (5.1) explicitly by $\tfrac{1}{6} c^i_{jkl}\bar{x}^j\bar{x}^k\bar{x}^l + O_4$ with the constants c^i_{jkl}, symmetric in j, k, l, to be found. One now goes on exactly as before, though the details are more tedious. The condition that the quadratic terms be absent leads to an explicit expression for c^i_{jkl} in terms of Γ^r_{pq} and its first derivative, all evaluated at $\overset{0}{x}{}^s$. But wait! We have so far ignored the identity $c^i_{j[kl]} = 0$. Explicitly, this requires that

$$\Gamma^i_{jl,k} - \Gamma^i_{jk,l} + \Gamma^m_{jl}\Gamma^i_{km} - \Gamma^m_{jk}\Gamma^i_{lm} = 0. \tag{5.4}$$

The function on the left—we shall meet it again—is of course evaluated at $\overset{0}{x}{}^r$. This, then, is the necessary and sufficient condition for terms quadratic in $\bar{x}^k$ to be removable from $\bar{g}_{ij}$, that is, for Euclidicity to obtain in an extended region. As we have seen, the linear terms can, however, always be removed.

Suppose them to have been so removed after a definite a^i_j has been chosen. Then one has a local inertial frame $\mathcal{I}$, in terms of which the coordinate system x^i may locally be directly interpreted. In other words, one will now be fully informed about the spatiotemporal relationships between events coordinatized by—"referred to"—any nonlocal system x^i which are local with respect to the event $\overset{0}{x}{}^i$. That is a step forward in the right direction; but we are still in the position of being able to describe only disjoint fragments of the overall histories of given particles.

There is a small gap which needs to be plugged. Recall that we agreed to regard any kind of material body notionally as being made up of elementary parts to which particlelike character is to be ascribed. Any such part, Q, will also in general be referred to some nonlocal coordinate system x^i which receives its local interpretation in terms of an appropriate inertial frame $\mathcal{I}$. That Q is not "free" is irrelevant, but what may seem puzzling is what one could possibly understand $\mathcal{I}$ to be here. $\mathcal{I}$ is supposed to consist of free particles, and how can one have these inside a material substance? I think there are two ways in which we might deal with this puzzle. One involves the time-honored method—already familiar from the study of electrostatics—of imaginary suitable cavities to be embedded within the substance for the sole purpose of being a habitat for $\mathcal{G}$. Questions of junction conditions arise, but let us simply take these to be resolvable. The second way is to look upon $\mathcal{G}$ from an "as-if" standpoint, that is, we still interpret the nonlocal coordinate system locally by reference to $\mathcal{G}$, merely pretending that the latter is physically realizable.

Velocities and similar kinematic concepts are part of the language of changing spatiotemporal relationships. By way of putting (5.1) to some illustrative use, contemplate the motion of a free particle near $\overset{0}{x}{}^{i}$, this event being supposed to occur in its history. If ds is the usual invariant interval between neighboring events in its history, its velocity is $d\bar{x}^{i}/ds$, and its acceleration $d^{2}\bar{x}^{i}/ds^{2}$ vanishes. Differentiating (5.1) twice, one finds directly that at $x^{i}=\overset{0}{x}{}^{i}$

$$\left(\frac{d^{2}x^{i}}{ds^{2}}\right)_{0}+\overset{0}{\Gamma}{}^{i}{}_{jk}\left(\frac{dx^{j}}{ds}\right)_{0}\left(\frac{dx^{k}}{ds}\right)_{0}=0. \tag{5.5}$$

Tempting though it may be, I shall not take the step of arguing as follows. Since the event $\overset{0}{x}{}^{i}$ may be thought of as any event of the particle's history, the index 0 may simply be removed from (5.5), the history as a whole consequently being characterized by the resulting differential equation. This equation in turn expresses the stationarity of the integral $\int ds$ with respect to arbitrary infinitesimal variations of the history between fixed arbitrary terminal events. In formal terms we would then say "the history of a free particle is a timelike geodesic"—timelike because $ds^{2}>0$ for any material particle. Well, although I do not reject this "geodesic principle" I much prefer not to adopt it at this point, for there are important issues at stake. Rather let us pretend for the time being that I haven't even mentioned it. Sooner or later we shall encounter it again.

In concluding this lecture let me draw your attention to this: when I spoke just now of velocity, and more particularly of acceleration, I was careful to refer only to quantities which were well-defined in terms of coordinates adapted to local inertial frames, that is to say, to normal coordinates. They are the quantities we know from special-relativistic kinematics. One has to talk about "acceleration" in particular with proper circumspection; and I believe that the use of terms such as "coordinate acceleration"—meaning $d^{2}x^{i}/ds^{2}$— should be avoided because of their misleading connotations. At any rate, it is not unusual to come across statements of the so-called weak principle of equivalence in a form something like this: "at a given height above the earth's surface all free particles sufficiently close to each other have the same acceleration." How can that be? Is it not the case that free particles are never accelerated? If a motion is accelerated it is not standard, but standard motion is *defined* by free particles. In short, within the theoretical framework within which we are operating the sentence "free particles are not accelerated" is analytic. Any assertion that a free particle has such and such nonzero acceleration simply reveals a confusion between different languages, namely, relativistic and prerelativistic languages; and by "relativistic" I now mean general-relativistic. In fact, we shall meet this kind of confusion again more than once.

Straight away one will ask this question: granted that in prerelativistic language the weak principle of equivalence seems to be a legitimate empirical law, have we somehow managed to avoid it altogether? Of course not. We took it as an observational fact that one could always arrange two particles to be relatively fixed at a mutual, sufficiently small distance $\frac{1}{2}\tau$ for a certain time $\tau^* \gg \tau$. The assertion of this possibility is an appropriate form of the weak principle of equivalence, consistent with the use of relativistic language. I do not think that the vagueness inherent in the unspecified time lapses τ and τ^* is a serious handicap because some such vagueness is implicit also in any other form of the principle. Be that as it may, without it the construction of local inertial frames which we undertook would not have been possible.

Perhaps I should add a word here about the "strong principle of equivalence." This, too, we have already taken into account by accepting that the local form of the equations governing basic physical phenomena—the field equations of electrodynamics for instance—do not allow any generic distinction to be drawn between any local inertial frames whatever. To all intents and purposes this *is* the strong principle of equivalence. It turns out, that it, too, contains an inherent fuzziness, but this is of a kind which we can examine, and perhaps dispose of, only later on. We shall be reminded of it near the end of the seventh lecture, for instance.

LECTURE 6

Evidently we now have one paramount need: to devise some strategy which will allow us to find the actual form of the functions g_{ij}. But wait!—is this not a dangerous way of stating the problem that confronts us? In the first place it won't do to speak here blandly of "the actual form" because given one such set of functions g_{ij}, any other set $g_{i'j'}$ obtained from it by an arbitrary transformation of coordinates will serve the same purpose and is therefore equally legitimate. In the course of obtaining a particular set g_{ij} one must at some stage necessarily impose a condition—a so-called coordinate condition—which temporarily inhibits the freedom to make arbitrary coordinate transformations. The second objection is this: the phrase "to find the actual form" does not make it crystal clear from the outset that, even granted the imposition of coordinate conditions, it is not a case of finding the g_{ij} once and for all, in the sense in which we initially found every local inertial frame to be characterized by the metric η_{ij}. In other words, whereas η_{ij} is absolute[II], g_{ij} is not. I ask: what leads us to make this assertion and just how is it to be understood?

For two closely connected reasons we must exercise particular care in answering this question. First, because, as I hinted earlier, the use of the term "absolute[II]" requires a cautionary comment, and this must now be provided. Second, because just at this point several distinct notions underlying the general theory of relativity come together to form a coherent whole. As regards "absolute[II]" the point at issue is this: it will have its intended meaning in a given context only provided one is using some kind of "space plus time language" not spacetime language. In other words, one has to be thinking in terms of the description of the temporal evolution of systems. I am, in fact, harking back to an earlier remark concerning the linguistic confusion implicit in phrases such as "the motion of a particle along its world-line". The world-line is a representative of a particle's history—how can one meaningfully say that the particle is "moving along it"? At any rate, suppose for the sake of argument that we have a way of coordinatizing any knowable event and, further, pretend that we somehow know the functions g_{ij} explicitly. This would amount to our having a knowledge of the history of the world—or perhaps you prefer the word "universe." I think it would then be legitimate to say that the g_{ij} are absolute[II]. I have set up this straw man only to shoot him down. The fact is that we do not know the g_{ij}, for whereas any particular observer may have knowledge of past events in the sense that he may have kept a record of certain observational results obtained in the past, he has no knowledge of any events which may occur in his future. Here, of course, I do not count as "knowledge" any predictive inferences which he may have drawn from laws and theories available to him. Therefore, to define the functions g_{ij} the results of observations actually made in the past must be combined with equations which describe their evolution, supplemented with initial and boundary conditions. Furthermore, these equations cannot be autonomous, that is to say, in general they cannot merely assert the vanishing of some functions of the g_{ij} and their derivatives.

What I am saying is that the metric tensor cannot be absolute[II]. To see this one need only reflect as follows. Quite simple observations reveal that the period τ^* over which any local inertial frame effectively maintains its integrity depends on the circumstances in which it happens to find itself, that is to say, depends on the spatiotemporal relations between it and other matter. By way of example, the value of τ^* is less for a frame near the earth —this "nearness" can easily be translated into the time-lapse language— than for one remote from it and in any case depends on the position of the moon. Now the differential equations contained in physical theories in themselves cover situations not only as they actually are, but also as they might be—only boundary conditions and the like nail down the actual state of affairs. From this point of view the position of the moon is, so to speak, "accidental." Quite generally, then, τ^* will depend on these accidental circumstances and so, in turn, will the g_{ij}. This last inference is based on the earlier transformation to normal coordinates, with the generic result that

$$\bar{g}_{ij}(\bar{x}^k) = \eta_{ij} + c_{ijkl}\bar{x}^k\bar{x}^l + O_3, \qquad (6.1)$$

the c_{ijkl} being functions of the g_{mn} and their first and second derivatives, evaluated at $\overset{0}{x}{}^l$. This local frame is therefore inertial in the proper sense only when the effects of the second term on the right of (6.1) are so small as to be negligible. The condition will cease to be satisfied when the $\bar{x}^i$ and in particular $\bar{x}^4$, exceed certain values depending on the values of the g_{ij} and their first and second derivatives at $\overset{0}{x}{}^k$. The conclusions which we have reached so far may now be summed up as follows: the metric tensor g_{ij} is not absolute[II]. It must satisfy differential equations which ensure consistency with—establish a lawlike connection between it and—the spacetime distribution of matter and energy.

One or two loose ends need to be tied up before I go on. First, whereas I argued that g_{ij} cannot be absolute[II] by drawing on rudimentary observations involving matter, I ended up talking about matter and energy. My warrant for doing so is, in essence, this: phenomenologically one cannot distinguish between them when they are jointly contained in a small, impermeable, adiabatic "black box." That has some rough edges, but I believe we may safely ignore them. Second, what justified me in surreptitiously passing from "equations" to "differential equations"? They cannot be algebraic if the g_{ij} are not to be absolute[II] and must therefore involve other fields derived from them by differentiation or integration. Here integration is, however, problematic since the nonconstancy of the $L^i_{i'}$ forbids the covariant addition of tensors attached to different points; and we would be faced with the need to introduce two-point tensors. In any event, the field equations of classical theories in the first instance usually take the form of differential equations, perhaps just because they can accommodate the accidental features of physical situations which I discussed a short while ago. New theories, however different from the old, do not evolve in isolation from them, and we are inclined to retain familiar features until we are forced to do otherwise.

Having disposed of these side issues we can proceed to complete the circle of the central argument. To begin with, imagine that a solution of the differential equations for the g_{ij}—provisionally I shall call them the "field equations" for short—corresponding to specified conditions, has been obtained. Of course, on the way the appropriate ancillary conditions—coordinate, boundary and initial conditions—will have been imposed. The specific functions g_{ij} now at hand govern the spatiotemporal relations between all events of that part of spacetime to which the solution refers. Now reflect! Whatever particles may be present, either individually or nominally as parts of material substances, their histories are themselves sequences of events. In other words, reverting to the language of temporal evolution, the way in which the spatial relations between particles evolve is known; and this is only a way of saying—more picturesquely, if less precisely—that we know the actual motion of all particles present, "particles" here being a catchall for material substances and the fields, such as the electromagnetic field, which mediate their mutual interactions.

The conclusion at which we have now arrived is crucial if the whole scheme is to hang together. Previously we had been able to describe fragmentary histories by referring them to local inertial frames, but were faced with the hiatus of being incapable of piecing them together. In particular, we knew that light rays could mutually connect pairs of particles even when they were not local with respect to each other—without this possibility even the schematic proposal for coordinatization which we contemplated earlier would have been empty. Yet we have no means of predicting what other particles any such ray might encounter along the way. In a predictive context causal connectibility over a nonlocal region to this extent remained an empty shell. Now, however, it seems that the hiatus is removed: we can describe finite histories.

Here a worrying thought crosses our minds. It seems that we have miraculously managed to avoid "equations of motion" altogether. How can that be? Equations of motion are the usual instrument for the construction of finite histories. We "integrate" them; and here, too, we have a kind of "piecing together," forces being connected locally with kinematical quantities. Is that not where we came in? Surely we had equations of motion in mind when we set out to establish the idea of a standard motion so that forces could be made responsible for any deviation from the standard?

We did, but there is no mystery. Let me draw upon an analogy, ignoring any misleading features it may contain. The history of a particle system of ordinary classical mechanics may be obtained by finding Hamilton's principal function by constructing, according to standard prescriptions, an appropriate integral of the Hamilton-Jacobi equation. In so doing no need arises for any explicit reference to equations of motion or their solutions. Is one now to say that one has managed without equations of motion and all that this phrase implies? Of course not, for the construction of the Hamiltonian which enters into the Hamilton-Jacobi equation is contingent upon the prior definition of standard motion, forces, and the like.

Pursuit of this analogy takes us one step further. One can show by mere differentiation that the equations between the coordinates, momenta, and the time which the principal function generates in fact satisfy the canonical equations of motion. If we are on the right track the assertion that a metric tensor g_{ij} is an appropriate solution of the field equations should—at least on an approximative level—be translatable into statements which assert that the coordinates of some identifiable point of any material body within the spacetime region under consideration satisfy ordinary differential equations of the kind formally recognizable as equations of motion. Nevertheless I believe that this is saying too much for two reasons: first, because whatever the appearance of such equations might happen to be, one would have to accord to them the status of equations of motion; second, strictly speaking such equations of motion are in principle redundant in any case. All we must ask for is that the metric tensor correctly represent what is

observed. In saying this I am including the temporal behavior of electromagnetic and other fields which enter into the description of interactions
through forces.

Let me briefly illustrate these remarks by a specific example, even if the
language used is a little ragged. Contemplate a solution g_{ij} of the field
equations which can be interpreted as belonging to a spacetime the only
actual events of which are the histories of two separated bodies P, Q. These
are subject to the following conditions: (1) P and Q are mutually fixed, (2)
both are symmetric about a light ray joining them, and (3) any particle
projected from P along such a ray will necessarily encounter Q. I then say
that P and Q are fixed. It is just here that we are being rather rough, for
strictly speaking the g_{ij} themselves, that is, the solution of the field equations, ought to account not merely for P and Q but also any incidental light
rays and particles to which we might appeal. I ignore this complication
provisionally: we shall have to return to it later. At any rate, there must in
fact be no solution having the proposed character for, as we know very well
from experience, P and Q will not remain permanently fixed. The equations
of motion which we are contemplating must certainly imply that this is so,
quite apart from whatever additional quantitative information they may
provide.

Though we have now laid its groundwork, the theory we are seeking to
establish will remain a mere shadow until such time as we succeed in
displaying the actual field equations in their final form. To find them is
therefore the immediate task before us and we turn out attention to it. It
goes almost without saying that we cannot deduce these equations. This is
certainly a matter of fact, if not of principle. One is reminded of the
equations of the electromagnetic field. Within the theoretical climate of his
day Maxwell could not obtain them by deductive reasoning and in fact had
to introduce the displacement current ad hoc because for various reasons it
seemed sensible to do so. What sensible steps, then, can we take to achieve
the desired end? Preferably we should make do with as few ad hoc
assumptions as possible, especially where they might be regarded as being
artificial and tailored merely to the case in hand. Under these circumstances
success will justify a sense of assurance that the equations at which we will
have arrived have to a large degree arisen from inner necessity and that they
alone are acceptable unless we are prepared to acquiesce in endless unwar-
ranted complications. Of course such feelings of assurance are not hard and
fast—they have an emotional element in them. Let us not, however, be
under any illusion that emotional motivations are not part of the stuff from
which progress in theoretical physics has often sprung in the past and will
surely do so in the future.

Be that as it may, we should prudently cast our minds back over what we have already done. Perhaps it contains material the full consequences of which remain yet to be explored? It does: about half way through the last lecture we took note of the fact that no coordinate system which one might introduce initially could be given preference over any other. *If* the g_{ij} were known functions one might, purely as a matter of convenience, prefer one coordinate system over another; but they are not. It is worth bringing to mind here once again that a coordinate system is merely a prescription for coordinatizing events. To be specific, let it be that proposed previously, that is, the schematic exchange of light pulses between a given event P and four other events Q_i. There the time lapses τ_i served as coordinates. Now, in place of the set Q_i pick some alternative set of events Q'_i as reference system, the choice of this being again largely arbitrary. The corresponding time lapses τ'_i will serve equally well as coordinates of P. τ_i and τ'_i thus represent two alternative coordinatizations of events, that is, we have two coordinate systems in hand and the generic relation $\tau'_i = \sigma_i(\tau_j)$ is a coordinate transformation. Each coordinate system has here a direct physical interpretation and in *this* case, moreover, the coordinate transformation represents a transition from one reference system to another. It would, however, be wrong universally to interpret the phrase "coordinate transformation" in this sense. A coordinate transformation is the transition from one coordinatization to another, however this transition may be contrived. In particular, it may be a mere mathematical device; and it is not legitimate to assume that for every "transformed coordinate system" one could find a reference system which will allow a straightforward time-lapse interpretation of the coordinates. In short, a coordinate transformation may be, but need not be, the representative of the switch from one reference to another.

We are now in a position to put forward an important proposition concerning necessary form invariance. To put it into words, I first introduce the new term "absolute form invariance." This is to be understood as follows. Absolute form invariance means proper form invariance when it is not contrived by the introduction of extraneous, absolute[II] objects. The proposition in question is now this: the equations which enter into the description of physical phenomena—and this includes those of immediate concern—must be absolutely form invariant under arbitrary transformations of coordinates. The absence of any privileged reference systems is certainly accommodated. Further, as we noted a moment ago, the class $\mathfrak{T}^*$ of coordinate transformations induced by transitions between alternative reference systems contains a high degree of arbitrariness but—presumably —does not exhaust the class $\mathfrak{T}$ of all possible coordinate transformations, restricted only by conditions of continuity and differentiability. However, I know of no effective way in which one might impose form invariance under $\mathfrak{T}^*$ other than by requiring form invariance under $\mathfrak{T}$.

Let us be clear that the proper form invariance of given equations usually obtains only because of the presence of one or more absolute[II] objects and one might introduce such objects merely for the purpose of being able to satisfy some demand for proper form invariance, as we have noted before. If one now insists that all such absolute[II] objects be absent one is clearly placing a powerful restriction on the possible form which equations, or sets of equations, may take. It may happen in the presence of some ancillary conditions either that the demand for absolute form invariance cannot be satisfied at all or else that the form of the equations is uniquely determined. In the latter case, one must take the view that it has definite "physical consequences" for, being part of a physical theory, the equations play an integral role in the description of physical processes. In saying this I am of course taking it for granted that one is not going to the semantic extreme of maintaining that no equation ever has physical consequences.

The requirement of absolute form invariance is intended to be a more carefully worded substitute for what is generally referred to as the "principle of covariance." This is, however, normally formulated so loosely and ambiguously that it has given rise to much misunderstanding and confusion. The more common formulations draw no distinction between proper and improper form invariance, nor do they explicitly require the absence of absolute[II] objects. At the lowest level the principle of covariance is then thought merely to amount to a demand for the tensorial character of equations—though why spinor equations should be so lightly dismissed I don't know. Again, as I already mentioned during the second lecture, it is said over and over again that the principle of covariance is "physically empty." Indeed, this view was put forward as early as 1917 on the grounds that any equation could be made "covariant." Curiously, Einstein himself concurred with this, having been led to believe that the principle served merely a heuristic function. Presumably he had only improper form invariance—the required tensorial character of equations—in mind. In any event, the opinion he expressed must be seen against the background of the conflict of so much concern to him, namely that between the relational and the absolute[III] views of spacetime. Still, it won't do.

Since we shall have occasion to return to these ideas later on, I shall not rehearse the argument in detail. Very briefly, "covariance" must here be interpreted as proper and not as improper form invariance; and such form invariance must obtain in the absence of absolute[II] objects. Concomitantly, one cannot make an equation covariant—meaning now absolutely form invariant—by "writing it in any one coordinate system, and then working out what it looks like in other arbitrary coordinate systems" for the simple reason that the first system will survive in the transformed equations as an absolute^{II} object. It remains to eradicate another incidental source of confusion. Equations involving η_{ij} may of course be written in general tensorial form. The place of the η_{ij} is then taken by the functions g_{ij}—and their derivatives—and there appears an additional covariant, that is, tensorial equation, for these. This equation is autonomous in the sense that it involves the g_{ij} and their derivatives alone and takes no cognizance of accidental situations as understood before. It therefore does not govern a temporal evolution within the proper meaning of the term and g_{ij}, like η_{ij}, is absolute^{II}. This is hardly surprising, since the equation in question merely declares that there exists a transformation of coordinates such that $g_{i'j'} = \eta_{i'j'}$.

To proceed, recall that special-relativistic theories were governed by an overriding regulative principle. In the case of Maxwell's equations, for instance, it turned out that these were already in harmony with it as they stood. Not so, however, the equation of classical mechanics. To achieve harmony they had to be modified to allow four-dimensional tensor equations to be written down which merely reproduced the original nonrelativistic equations in the special case of sufficiently slow motions. In other words, the factual utility of the regulative principle was this: it supplied a straightforward test for ascertaining whether given equations were in harmony with the special principle of relativity or not. Now we have once again come face to face with a powerful regulative principle. Yet the situation differs greatly from that just described in as far as we have no equations for the g_{ij} which we might submit to a test. We have no choice but to investigate to what extent the principle, when coupled with ancillary assumptions, will so circumscribe the admissibility of equations that one is effectively led to a unique set.

Regulative principles are always powerful since they act as selection rules among whole classes of theories. This strongly suggests that we should propose a second regulative principle to stand alongside the first. When one looks back over existing theories it soon becomes obvious that there is not much to choose from. Common to many of them, however, is the presence of variational principles, in the sense that their basic equations, whether field equations or not, are expressions of the stationarity of some action functional V: in other words, they express the vanishing of the functional derivative of some Lagrangian L. The ubiquity of variational principles is accounted for by the fact that they imply the existence of certain differential identities which can moreover be easily discovered—think of Nöther's theorem. That these identities find an interpretation as theorems of conservation is immaterial. What is important is that they exist at all. This certainly suggests that here, too, one should insist on a formulation of the required equations which leans heavily on functional derivatives. After all, we already know that these equations must in their very nature make reference to distributions of matter and energy. These are subject to certain lawlike restrictions; equivalently, they are subject to "equations of motion." It is natural to foresee the possibility that the differential identities just entail these equations of motion.

We cannot expect that the two regulative principles, taken together, will by themselves lead to a unique set of field equations. Granted the second, the first is not likely to select a unique Lagrangian; on the contrary there will be several, perhaps an unlimited number of them. Of course we could simply pick out some sequence from among these, more or less at random, set out the resulting theory in detail each time, and submit its predictions to observational test. Clearly this is not a practical proposal. What one needs is yet another regulative principle for the purpose of further restricting the selection of admissible Lagrangians. Here we naturally turn to a vague stock-in-trade of the natural scientist, namely an appeal to the "principle of simplicity." It is difficult to discover what this is supposed to be. Sometimes it is an invocation of "Occam's razor": if the introduction of some concept, entity, or construct into a formal description is not strictly necessary then it is to be avoided altogether. It then operates as a principle of economy or parsimony. At other times its intended meaning is even more elusive. First there is the question of simplicity of a theory on various levels, the level of its formal calculus, the level of the observational interpretation of this, and, by no means least, the computational level. How is one to attach relative weights to these? To what extent is it all in the end just a matter of practical convenience? There are other aspects into which the notion of simplicity enters, virtually on a psychological level: it is occasionally held that the theory is "simpler" than another because it is the first which most closely resembles—in structure and concept formation—some third theory which happens to be hallowed by tradition. All this is further bedeviled by a crucial factor: how is "simplicity" to be quantified? Can one give some number which is a measure of the "degree of simplicity"? In general, surely not. Fortunately the issue is fairly clear-cut under special circumstances. For instance, it will be universally agreed that a linear algebraic equation is simpler than one that is quadratic. Here the least number of instructions to be supplied to a pocket calculator for the purpose of solving one equation or the other actually provides a hard and fast measure of simplicity, granted of course that there is such a least number. Again, an ordinary differential equation of order two is simpler than one of order four in the sense that the first requires for a specific solution the prescription of fewer initial conditions than the second. Yet here the issue is already clouded: would one really want to call the equation $y'' + (x^2 - 1)y = 0$ simpler than the equations $y'''' = 0$? If not, what criterion of simplicity has been substituted? Finally, when one succeeds in formulating a valid theory which is simple by common consent, does this reflect some kind of "simplicity of nature"; or is its simplicity perhaps merely a reflection of a well-designed methodology?

We may here allow ourselves the luxury of leaving these questions unresolved, since fortunately we are not concerned at the moment with forming a judgment about the simplicity of a theory as a whole. Within a much more restricted compass we simply treat the principle of simplicity as a selection rule for a particular Lagrangian, granted that we shall be able to justify the assertion that such and such a Lagrangian is in fact the simplest among those which could possibly be admitted.

LECTURE 7

By the time we reached the end of the preceding lecture we had adopted, after lengthy discussion, the following tentative view: the differential equations which the metric tensor g_{ij} has to satisfy is to be discovered by following a course largely determined by three distinct guidelines. They are:

P1: the principle of absolute form invariance,
P2: the prescribed presence of functional derivatives,
P3: the principle of simplicity.

The character of P2 and P3, especially the latter, is heuristic, but that of P1 is not.

Let us see whether we are on the right track. To this end it is advisable first to consider a part of some spacetime void of matter and energy. The g_{ij} are then the only field functions present, "field" being understood as a technical term. Then P2 is accommodated by requiring the field equations to have the generic form

$$P^{ij} := \frac{\delta L}{\delta g_{ij}} = 0, \qquad (7.1)$$

where L is an as yet unknown Lagrangian. (7.1) expresses the stationarity of

$$V := \int L(-g)^{1/2} d^4x \qquad (7.2)$$

with $g := \det g_{ij}$, as usual. Because of P1, L must certainly be a scalar, so that for a given boundary B of the region of integration V is simply a number. Let h_{ij} be an arbitrary infinitesimal variation of the g_{ij}. There will be a corresponding variation δV of V. When the h_{ij} and their successive derivatives to a certain order vanish on B,

$$\delta V = \int P^{ij} h_{ij} (-g)^{1/2} d^4x, \qquad (7.3)$$

and this in fact defines the symmetric covariant tensor P^{ij}. As a particular case of this, the variation h_{ij} may be merely a consequence of an infinitesimal transformation of coordinates

$$x^k = \tilde{x}^k + \xi^k(\tilde{x}^l) \qquad (7.4)$$

which vanishes on B. I remark in passing that the use of the kernel-index notation would here lead to needless complications. Bearing in mind that g_{ij} is a tensor and that x^k and $\tilde{x}^l$ are coordinates of the same point, one convinces oneself easily that

$$h_{ij} = g_{ij,k}\xi^k + g_{ik}\xi^k{}_{,j} + g_{jk}\xi^k{}_{,i}. \qquad (7.5)$$

The right hand member of this does not manifestly have the appearance of a covariant tensor. A small digression on notation seems unavoidable if we are not to be handicapped by needless formal complexities later on.

Whereas ξ^k is a vector field its derivative $\xi^k{}_{,l}$ is, in general, not a tensor field, since one is in effect mutually subtracting tensors attached to different points x^l and $x^l + dx^l$. In fact,

$$\xi^k{}_{,l} = L^k{}_{k'} L^{l'}{}_l \xi^{k'}{}_{,l'} + L^k{}_{k',l'} L^{l'}{}_l \xi^{k'}. \qquad (7.6)$$

This shows explicitly that $\xi^k{}_{,l}$ is not a tensor under general transformations of coordinates. It is therefore natural to define a differential concomitant $\xi^k{}_{;l}$ of ξ^k as follows:

$$\xi^k{}_{;l} = \xi^k{}_{,l} + \Gamma^k{}_{lm}\xi^m, \qquad (7.7)$$

where $\Gamma^k{}_{lm}$ — the so-called coefficients of linear connection — are in the first instance arbitrary except in as far as they are prescribed to obey the law of transformation

$$\Gamma^{k'}{}_{l'm'} = L^{k'}{}_k L^l{}_{l'} L^m{}_{m'} \Gamma^k{}_{lm} + L^{k'}{}_k L^k{}_{l',m'}. \qquad (7.8)$$

This condition suffices to ensure that the right hand member of (7.7) is a tensor. However, we are not allowed to introduce into our work any new absolute[II] quantity. We conclude that Γ^k_{lm} must be a function of g_{ij} and its derivatives alone. However, only by imposing an ancillary condition can we fix the form of Γ^k_{lm} uniquely. The condition is this: whenever ξ^k and its concomitants are referred to a normal coordinate system at any event P, then $\xi^k_{;l}$ shall at P reduce to $\xi^k_{,l}$, that is, the Γ^k_{lm} must vanish. From this we infer that the Γ^k_{lm} are in fact Christoffel symbols. The latter certainly vanish because in a normal coordinate system the $g_{kl,m}$ do so and one confirms by direct transformation that they obey (7.8). Moreover, suppose there existed another linear connection $\bar{\Gamma}^k_{lm}$ satisfying all our requirements. Bearing in mind that according to (7.8) the difference between two linear connections is always a tensor, one would then have a tensor which vanishes at any point P when referred to normal coordinates there; but a tensor which vanishes in one system vanishes in all systems. Since P is arbitrary this implies that Γ^k_{lm} is uniquely fixed by the conditions which have been imposed.

$\xi^k_{;l}$ is called the "covariant derivative" of ξ^k. The following elementary remark may be illuminating. In the context of nonrelativistic theories one is accustomed to using Cartesian coordinates. Occasionally, however, one goes over to arbitrary curvilinear coordinates. What are the formal effects of doing so? If suffices to illustrate these by investigating the derivative $\xi^k_{,l}$, supposed to relate to Cartesian axes. An arbitrary transformation of coordinates takes this immediately into the right hand member of (7.6). $L^k_{k',l'}$ may be removed by means of (7.8), taking into account that Γ^k_{lm} of course vanishes, and one ends up with $L^{l'}_l L^k_{k'} \xi^k_{;l'}$. The covariant derivative has, so to speak, appeared of its own volition.

The notion of covariant derivative, or covariant differentiation, can be extended to the general tensor density. If $T^{ij\cdots}_{kl\cdots}$ is a tensor, or more generally a tensor density of weight w, then its covariant derivative is

$$T^{ij\cdots}_{kl\cdots;m} := T^{ij\cdots}_{kl\cdots,m} + \Gamma^i_{pm}T^{pj\cdots}_{kl\cdots} + \Gamma^j_{pm}T^{ip\cdots}_{kl\cdots}$$

$$+ \cdots - \Gamma^p_{km}T^{ij\cdots}_{pl\cdots} - \Gamma^p_{lm}T^{ij\cdots}_{kp\cdots} - \cdots$$

$$- w\Gamma^p_{pm}T^{ij\cdots}_{kl\cdots}. \tag{7.9}$$

The covariant derivative of a tensor evidently depends linearly on the tensor. Further, covariant differentiation obeys Leibniz' rule. All indices following a semicolon imply covariant differentiation.

One observes immediately that the covariant derivative of g_{ij} vanishes identically; one says that "g_{ij} is covariant constant." This brings with it that the operations of covariant differentiation and of juggling indices mutually commute; and that is a matter of enormous convenience. In special cases the $\Gamma^k{}_{lm}$ will be absent from tensors nominally involving covariant derivatives: by inspection $T_{[kl...;m]} \equiv T_{[kl...,m]}$. In particular, if u_i is any vector field, $u_{[i;j]} \equiv u_{[i,j]}$.

One can now write (7.5) in the manifestly covariant form

$$h_{ij} = 2\xi_{(i;j)}. \tag{7.10}$$

(7.3) becomes

$$\delta V = 2 \int P^{ij}\left(\xi_{i,j} - \Gamma^k{}_{ij}\xi_k\right)(-g)^{1/2}d^4x.$$

The first term on the right may be integrated by parts, the surface integral being rejected since ξ_i vanishes on the boundary. Upon then replacing ordinary by covariant derivatives one not unexpectedly finds that

$$\delta V = -2 \int P^{ij}{}_{;j}\xi_i(-g)^{1/2}d^4x.$$

Now recall that V is a number obtained by invariantly integrating a scalar field over a fixed region with boundary B. This number cannot depend on the choice of coordinate system within B. One concludes that δV must vanish. However, B was after all arbitrarily chosen, and may so lie within an arbitrarily small neighborhood of any chosen point. We conclude that

$$P^{ij}{}_{;j} = 0; \tag{7.11}$$

that is, the covariant divergence of P^{ij} vanishes identically. One sometimes says that "P^{ij} is conserved." It is a convenient phrase, but it should not be taken to mean more than that which it replaces. Our first aim is now attained: any Lagrangian, as long as it is invariant, generates a functional derivative which is subject to a certain covariant differential identity.

The choice of a specific Lagrangian can no longer be deferred. Of course, in speaking of "choice" I do not mean to imply that L is to be "taken out of a hat." Rather, the extent to which $P1$ and $P3$ in fact dictate the actual form of L must be investigated. To begin with, L must be a function of the g_{ij} and its successive derivatives alone, for $P1$ demands the absence of any extraneous objects. L cannot depend explicitly on the coordinates since any such dependence would certainly entail violation of $P1$. In short, L must be an invariant function of the g_{ij}, the $\Gamma^m{}_{kl}$ and of the successive derivatives of the latter. Here I have replaced the first derivatives of the g_{ij} by Christoffel symbols. This step is legitimate since they uniquely determine each other. In the light of P3 we now contemplate placing upon L the following restrictions in turn: L depends (1) on the g_{ij} alone, (2) on the g_{ij} and $\Gamma^m{}_{kl}$ alone, (3) on the g_{ij}, $\Gamma^m{}_{kl}$, and $\Gamma^s{}_{pq,r}$ alone, and so on, with higher and higher derivatives of the Christoffel symbols.

The first possibility as such is not viable for it leads to no differential equation. Nevertheless, there is an invariant function, albeit a trivial one, of the g_{ij}, namely $g_{ij}g^{ij}=4$; and, supplied with a constant factor, it can always be added to any other Lagrangian which we may come to regard as suitable. The second possibility can be excluded from the start. Any such invariant would be a function of invariant transvections of metric components and Christoffel symbols. These will, however, vanish in normal coordinates at any arbitrarily selected point and so the transvections in question will vanish everywhere. In other words, apart from a constant there is no invariant of the kind under consideration, and we must go on at least to the third alternative.

This task seems at first sight rather intimidating. Still, in the spirit of P3 we may initially impose the provisional restriction that L shall depend linearly on the derivatives of the $\Gamma^m{}_{kl}$. A nonlinear dependence would surely not be as "simple" and in any event this alternative, too, can be examined later. Now, to construct an invariant which contains the $\Gamma^s{}_{pq,r}$ linearly, any such derivative must be combined—transvected—with Christoffel symbols, metric components, and Kronecker deltas in such a way that only dummy indices remain. This is of course only a matter of necessity, not of sufficiency. At least one such transvection must not contain any Christoffel symbol as factor for otherwise there cannot exist any invariant of the required kind. This follows from the same argument as that which previously disposed of the second possibility. Inspection reveals, moreover, that there are only two distinct transvections of the required kind. They are $Y:=g^{ij}\Gamma^k{}_{ik,j}$ and $Z:=g^{ij}\Gamma^k{}_{ij,k}$. Now go over to primed coordinates. Using (7.8), with primed and unprimed indices interchanged, one finds that the transform Y' of Y consists additively of terms only one of which, namely $g^{p'q'}L^i{}_{p'}L^m{}_{q'}L^j{}_{k'}L^{k'}{}_{i,jm}$, contains second derivatives of the $L^{s'}{}_t$. Exactly the same term is present in the transform Z' of Z. However L may be constituted within the limitations hitherto imposed upon it, the absence from its transform L' of the term just referred to is assured if and only if Y and Z occur in L in the combination $Y-Z$.

Among the terms which go to make up $Y'-Z'$ only one is free of Christoffel symbols, namely $g^{p'q'}L^i{}_{p'}L^l{}_{k'}(L^j{}_{q'}L^k{}_{l'}-L^k{}_{q'}L^j{}_{l'})L^{l'}{}_{kl}L^{k'}{}_{ij}$. By inspection, the disappearance of this from L' can be arranged by supplementing $Y-Z$ additively with the expression $C:=g^{ij}(\Gamma^m{}_{ik}\Gamma^k{}_{jm}-\Gamma^m{}_{ij}\Gamma^k{}_{km})$, for the terms of its transform which are free of Christoffel symbols are just such as to achieve the required cancellation. Then, however, it turns out that all the transformation coefficients and their derivatives disappear from $Y' - Z' + C'$. In other words, $Y - Z + C$ is a scalar. Moreover, one cannot construct any other scalar by the addition of further nonconstant terms since all of them will have Christoffel symbols as factors and therefore cannot collectively constitute a scalar. This follows as usual by introducing normal coordinates at any selected point.

What we have thus found by quite elementary means is this: to within a numerical factor the only Lagrangian which is absolutely form invariant, depends linearly on the first derivatives of the $\Gamma^k{}_{ij}$, and contains no higher derivatives of these is

$$L = g^{ij}\left(\Gamma^k{}_{ik,j} - \Gamma^k{}_{ij,k} + \Gamma^m{}_{ik}\Gamma^k{}_{jm} - \Gamma^m{}_{ij}\Gamma^k{}_{km}\right) \qquad (7.12)$$

apart from an additive arbitrary constant— -2λ, say—about which I shall have more to say. This particular scalar is usually represented by the symbol R and I call it the Ricci scalar. Further, one may confirm by direct transformation that the factor with which g^{ij} is transvected is itself a tensor. It is symmetric, bearing in mind that $\Gamma^k{}_{ik} = \frac{1}{2}(\ln|g|)_{,i}$. It is known as the Ricci tensor and denoted by R_{ij}. Incidentally, already during the fifth lecture we encountered the quantity

$$R^l{}_{ijk} := \Gamma^l{}_{ik,j} - \Gamma^l{}_{ij,k} + \Gamma^m{}_{ik}\Gamma^l{}_{jm} - \Gamma^m{}_{ij}\Gamma^l{}_{km}. \qquad (7.13)$$

Inspection reveals that R_{ij} is a contraction of this, namely $R^l{}_{ijl}$. We suspect that $R^l{}_{ijk}$ is a tensor. That it is indeed so may be confirmed by direct transformation. There is, however, a simpler way, as follows. Take an arbitrary vector field u_i and calculate the second covariant derivatives $u_{i;jk}$ and $u_{i;kj}$. There is no reason to assume that they are necessarily equal. Indeed, in general they are not. Explicit evaluation shows directly that

$$2u_{i;[jk]} = R^l{}_{ijk}u_l. \qquad (7.14)$$

Since the left hand member is a tensor and u_l an arbitrary vector the quotient theorem shows that $R^l{}_{ijk}$ is a tensor. It is known as the Riemann tensor. Evidently its vanishing is the necessary and sufficient condition for covariant differentiation to be a commutative operation. I shall postpone some further brief remarks concerning $R^l{}_{ijk}$ till the end of this lecture in order to avoid further disruption of this part of the argument and go on immediately to find the explicit form of the functional derivative of R. First, an infinitesimal variation h_{ij} of g_{ij} induces a variation $\gamma^k{}_{ij} := \delta\Gamma^k{}_{ij}$ of $\Gamma^k{}_{ij}$.

Since this is the difference between two linear connections it is a tensor. δR_{ij} is of course also a tensor. One therefore concludes by inspection that $\delta R_{ij} = 2\gamma^k{}_{i[k;j]}$, since the right hand member, by the usual argument, cannot contain additional terms all of which have Christoffel symbols as factors. Then, with $q := (-g)^{\frac{1}{2}}$,

$$\delta \int R q d^4 x = \int \left\{ 2q g^{ij} \gamma^k{}_{i[k;j]} + R_{ij} \delta (q g^{ij}) \right\} d^4 x.$$

The derivatives of the $\gamma^k{}_{il}$ which appear here may be removed by integrating by parts, the integrated terms being rejected since all variations are supposed to vanish on the boundary. As a result the first term of the integrand is to be replaced by $-2\gamma^k{}_{i[k}(q g^{ij})_{;j]}$, the absence of further terms being assured by the usual argument. This last expression, however, vanishes on account of the covariant constancy of g_{ij}. As regards the second term, since $\delta g^{ij} = -g^{im} g^{jn} \delta g_{mn}$ and $\delta q = \frac{1}{2} q g^{mn} \delta g_{mn}$, it becomes $(-R^{ij} + \frac{1}{2} g^{ij} R) q \delta g_{ij}$. It follows that the required functional derivative is

$$\delta R / \delta g_{ij} = \tfrac{1}{2} g^{ij} R - R^{ij}.$$

The tensor on the right is usually called the Einstein tensor and denoted by G^{ij}. We do not need to prove that it is conserved: we already know it to be so.

As a matter of principle, I must just rectify a small sin of omission though it happens to be inconsequential in the present context. When forming the quantities Y and Z, I appeared to imply that for the purposes of transvection only Kronecker deltas, metric components, and Christoffel symbols were available, though the last of these had to be rejected. In fact, however, one must include in the list also the alternating tensor $e^{ijkl} := q^{-1} \epsilon^{ijkl}$, where ϵ^{ijkl} is the Levi-Civita tensor density. It so happens that nothing is gained here. Because of the skew symmetry of e^{ijkl} any completely contracted product of $\Gamma^s{}_{pq,r}$, e^{ijkl} and metric components vanishes identically.

The Lagrangian $L = R - 2\lambda$ is "even simpler" than we might have expected. First, let the "order" of a function be the order of the highest derivative of the metric tensor which occurs in it. Then, if one wishes to have functional derivatives of at most the second order one would normally restrict one's investigation to Lagrangians of the first order, but here, as we saw, no acceptable Lagrangian of this kind can be constructed at all. A second order Lagrangian will, however, lead to fourth order functional derivatives, save under exceptional circumstances. It so happens that these obtain here, for $P^{ij} = G^{ij} - \lambda g^{ij}$, and this is manifestly of the second order.

Lagrangians which are nonlinear in the $\Gamma^k{}_{ij,l}$ necessarily lead to fourth order functional derivatives, at any rate in a four-dimensional context. When L is of third or higher order in which case, as an added complication, the highest derivatives must occur nonlinearly, the state of affairs is even more complicated. Actually, so-called non-linear Lagrangians are sometimes considered. These involve invariants of the Riemann tensor such as R^2, $R_{ij}R^{ij}$ and $R_{ijkl}R^{ijkl}$. The difficulties one then encounters are enormous, both conceptually and computationally. We may safely take all these possibilities as being at variance with $P3$.

Now reflect! Should we not also take the presence of the constant $\dot\lambda$ in L as being at variance with $P3$? I can make no compelling case for this view. Einstein first introduced λ, only to reject it later on the grounds that its presence would seriously reduce the logical simplicity of the theory and would also be aesthetically objectionable. In fact, he regarded the introduction of λ as "the greatest blunder of his life." It is not clear why he should have gone to such lengths. He seems to have regarded it in some way as being in conflict with Machian ideas, of which more later. Then, too, if the addition of -2λ to R—it was only an afterthought—had never occurred to him he might have been able to predict certain large scale features of the world which were later revealed by observation. However that may be, unless we are willing to be swayed by the mere voice of authority this does not help us, for it seems that one only has the invocation of $P3$ again. One might argue that unless $\lambda=0$ the equations $R_{ij}=\lambda g_{ij}$ do not admit the particular solution $g_{ij}=\eta_{ij}$. But why should they; provided, at any rate, that the value of λ is sufficiently small? Whatever other reasons might be advanced later for the exclusion of λ from L, I shall not rehearse them now, but return to them as occasion arises. For the present at least I shall continue to retain it.

Hitherto we have restricted ourselves to parts of spacetime void of matter and energy, that is, to "vacuum spacetime regions." Three regulative principles, taken alone, combined to lead to all intents and purposes uniquely to the vacuum equations $G^{ij}-\lambda g^{ij}=0$ or, equivalently, $R_{ij}=\lambda g_{ij}$. It remains to find general equations expressing the consistency between the g_{ij} and the distribution of matter and energy. That $P1$ must again operate is without question. As regards $P2$, we are inclined to adhere to this, too, as far as possible, whilst $P3$ may reasonably be taken to play a part in the formulation of any sort of physical theory.

If only for orientation, we shall do well to deal first with a special case, albeit a very important one: that of the Maxwell vacuum field. Recall the special relativistic equations (4.5). The first merely asserts, in any simply connected region, the existence of a vector field A_i such that $f_{ij}=-2A_{[i,j]}$.

Transcribed to general curvilinear coordinates the equations thus are in effect

$$f_{ij} = -2A_{[i,j]}, \qquad g^{jk}f_{ij;k} = j_i. \qquad (7.15)$$

Maxwell's theory is formally a vector theory, the first member of (7.15) functioning as a definition of f_{ij}. Then the second member expresses the vanishing of the functional derivative with respect to A_i of the Maxwell Lagrangian

$$M := \tfrac{1}{4}f_{ij}f^{ij} - A_i j^i, \qquad (7.16)$$

where j^i must be regarded as not subject to variation. The usual electromagnetic stress-energy-momentum tensor—energy tensor for short—is

$$T^{ij} = \tfrac{1}{4}g^{ij}f_{kl}f^{kl} - f^{ik}f^j{}_k. \qquad (7.17)$$

It is symmetric and its divergence is

$$T^{ij}{}_{;j} = f^{ij}j_j$$

as a consequence of the field equations (7.15). M is the sum of the pure field Lagrangian $M^* := \tfrac{1}{4}f_{ij}f^{ij}$ and a part which relates to currents and their interactions.

To be consistent we need to limit ourselves now to what I shall call the strict Maxwell vacuum field which is characterized by the absence of currents, for the following reason. To admit currents is not merely to admit charges but their material carriers as well. The term $A_i j^i$ of M, however, makes no reference to the material aspects of currents. Since for heuristic reasons we want to remain on the simplest level for the time being, we must take $j^i = 0$ in the spacetime region of interest.

Recall that the equation of consistency between the metric and the distribution of matter and energy is certainly dominated by $P1$. Here this means that it is an absolutely form invariant equation which contains no quantities other than the g_{ij}, the f_{kl}, and their derivatives. If we adhere to $P2$, the "Einstein-Maxwell Lagrangian" L must share these properties. Now, when we spoke of "equation of consistency" we had the equation $\delta L/\delta g_{ij} = 0$ in mind; yet it would be more natural to subsume under that phrase the joint equations $\delta L/\delta g_{ij} = 0$ and $\delta L/\delta A_k = 0$. I won't press the point, for reasons which will become plain later on. For the moment, at any rate, we will expect that, referred to a local inertial frame and curvilinear coordinates, the equations governing the Maxwell field should be $f^{ij}{}_{;j} = 0$ or at least closely resemble these. This last reservation has to be made because the g_{ij} are now not, in general, the transforms of η_{kl}. Nevertheless these equations amount to the vanishing of the functional derivative of M^*. In the spirit of $P3$—and also following the prescriptions of established theories—it suggests itself that

$$L = R - 2\lambda - 2\kappa M^* \qquad (7.19)$$

should be taken as the Einstein-Maxwell Lagrangian. Here κ is a fixed constant, yet to be chosen once and for all.

Before pursuing further the question of alternative possibilities let us investigate some implications of (7.19). That $\delta L/\delta A_i = 0$ is to be taken as the equation governing the Maxwell vacuum field we already know, but of course we have to bear in mind that the g_{ij} which occur in it are not given—absolute[II]—functions. Let us disregard the equation and go on to $\delta L/\delta g_{ij}$. To obtain $\delta M^*/\delta g_{ij}$ one has to vary M^*q with respect to g_{ij}, with f_{kl} fixed, of course. The work is quite elementary and yields the result $\delta M^*/\delta g_{ij} = \frac{1}{2}T^{ij}$. Then the equation of consistency $\delta L/\delta g_{ij} = 0$ is

$$G^{ij} - \lambda g^{ij} = \kappa T^{ij}. \qquad (7.20)$$

This is a very remarkable equation, for it requires that because the divergence of its left hand member vanishes, $T^{ij}_{\;\;;j}$ must also vanish. However, taking the definition of $f_{ij} := 2A_{[j;i]}$ into account—it entails that $f_{(ij)} = 0$ and $f_{[ij;k]} = 0$—one has

$$T^{ij}_{\;\;;j} = f^i_{\;k} f^{kj}_{\;\;;j},$$

and the vanishing of $T^{ij}_{\;\;;j}$ requires the vanishing of $f^{kj}_{\;\;;j}$, unless it so happens that $\det f^i_{\;k} = 0$, which in ordinary language expresses the mutual perpendicularity of **E** and **B**. One would hardly expect to come across fields which have this property exactly over a finite region. In fact, I do not know how to construct them on paper or even whether one can do so at all. Bear in mind that one cannot any longer think of Maxwell's equations as being linear, in the sense that the equations which govern it are coupled nonlinear equations for the g_{ij} and f_{kl} jointly. At any rate, we may safely disregard the possibility that $\det f^i_{\;k} = 0$ and so conclude that the equation $f^{ij}_{\;\;;j} = 0$ is a consequence of the equation of consistency (7.20). The equation which we chose to disregard has thus reappeared along another route. This looks like a pointer to the possibility that equations of motion may indeed be contained in the equation of consistency.

That is encouraging. Nevertheless we should briefly examine whether this justifies the mere juxtaposition by addition of $R - 2\lambda$ and M^*. It is true that T^{ij} describes the distribution of stresses, momentum, and energy in the electromagnetic field when $g_{ij} = \eta_{ij}$ and only under these circumstances must the equations of the field reduce to their customary form. All this, however, does not exclude the possibility that M^* might be a function of the g_{ij}, f_{kl} or their derivatives more general than that hitherto contemplated. Such greater generality can always be thought of as achieved by the addition of further terms to M^*, provided that these vanish when $g_{ij} = \eta_{ij}$. After all, to the extent that one can allow oneself to speak of the electromagnetic equations as a separate entity at all, those with which we are dealing are simply not the usual equations. Who knows what might be the case under extreme circumstances?

These are uncertainties over and above those which we disposed of by appealing to $P3$ in the context of deciding on the "pure" Lagrangian R. We have no option but to appeal to $P3$ again. How else are we to exclude the appearance of so relatively simple an additional term as $M_1 := R_{ijkl} f^{ij} f^{kl}$, for example? Other complications quite apart, this would induce in the functional derivatives of L the presence of further terms containing not only the covariant derivative of the Riemann tensor, but also the first and second covariant derivatives of f_{ij}. This increase in order alone is good reason for rejecting M_1. It is true that there exists the possible addition to M^* of the scalar $M_2 := \delta^{ijkl}_{mnst} R_{ij}{}^{mn} f_{kl} f^{st}$, which is unique in as far as it does not lead to the appearance of higher derivatives. However, written out in full, it is so complex that $P3$ demands its exclusion. You may have noticed that both M_1 and M_2 of course have the Riemann tensor as factor, so that they vanish when $g_{ij} = \eta_{ij}$. The requirement that terms of this kind be rejected from L may be called the "principle of minimal coupling," equation (7.20) being formally regarded as representing a coupling between g_{ij} and f_{kl}. I shall refer to it again in due course. Note, however, that it is just a special version of $P3$.

At the end of the fifth lecture I made some remarks about the strong principle of equivalence. In the present context it evidently implies the principle of minimal coupling, for if Maxwell's equations are understood to be $\delta L / \delta A_i = 0$, the additional terms generated by the inclusion of M_1 in L, for example, will not in general vanish in local inertial frames and to this extent these would then be distinguishable from each other. To what extent one regards this as additional justification for rejecting electromagnetic Lagrangians other than M^* is a matter for individual judgment, for we shall meet other circumstances in which the rejection of "additional terms" which fail to vanish in normal coordinates would be illegitimate.

Finally the promised brief remarks concerning the Riemann tensor. First, consider a Euclidean spacetime. If the coordinates are Cartesian all Christoffel symbols vanish and so R^l_{ijk} vanishes. Being a tensor, it therefore vanishes for every choice of coordinate system. The condition $R^l_{ijk} = 0$ for Euclidicity is, however, not merely necessary, but also sufficient. An instructive justification for this assertion goes back to the construction of normal coordinates undertaken in the fifth lecture. There we learned that the terms of $\bar{g}_{ij}$ quadratic in the $\bar{x}^k$ could be removed only if $R^l_{ijk} = 0$. Let this condition be satisfied. Then it turns out—we need not consider the details —that all terms of yet higher degree in the $\bar{x}^k$ can also be removed from $\bar{g}_{ij}$ by carrying on the process used before; in short, the metric has been constructively reduced on the form $\eta_{ij} d\bar{x}^i d\bar{x}^j$.

The Riemann tensor has a variety of important algebraic and differential properties. To survey them it is best to go over to its covariant concomitant:

$$R_{ijkl} = g_{j[k,l]i} - g_{i[k,l]j} + 2g_{mn}\Gamma^m_{i[l}\Gamma^n_{k]j}. \qquad (7.21)$$

From this one reads off the following symmetries:

$$R_{(ij)kl} = 0, \quad R_{ij(kl)} = 0, \quad R_{ijkl} = R_{klij}, \quad R_{i[jkl]} = 0. \qquad (7.22\text{a-d})$$

These reduce the number of linearly independent components to 20.

Now let P be an arbitrary point and contemplate the covariant derivative $R^l_{ijk;m}$ at P, referred to coordinates normal at P. One has $R^l_{ijk;m} = 2\Gamma^l_{i[k,j]m}$, so that by inspection

$$R^l_{i[jk;m]} = 0. \qquad (7.23)$$

This is a tensor equation, so that it holds generally. It is known as the identity of Bianchi. Transvection of (7.23) with $\delta^m_l g^{ij}$ leads at once to the conclusion that $G^m_{k;m} = 0$, but that merely confirms what we already know.

LECTURE 8

With the Einstein-Maxwell equations as a guide, we can now attempt to go on to quite general situations. Of course, we shall do so under the auspices of the three regulative principles P1–P3. Perhaps you will object that I am constantly giving too much weight to P3, this being little more than a heuristic guide—in effect, at any rate. The only alternative would be a more sustained appeal to observational results. However, when the aim is one of great generality one is virtually forced to have some kind of complete theory first whose predictions can then be tested by experiment: and it is just such a theory that we are trying to construct. Also, is it not a fact that within presentations of all manner of theories phrases such as "we make the simple assumption..." commonly abound? This is, of course, not so in the case of "axiomatic theories" but then one is merely presenting what is already established in supposedly final form, and the real perplexities are swept under the carpet. At any rate, a casual survey of results already achieved and of the arguments on which they rest lends compelling weight to the following assumption: the general equation of consistency has the generic form

$$P^{ij} := G^{ij} - \lambda g^{ij} = \kappa T^{ij}, \qquad (8.1)$$

where T^{ij} is now a symmetric energy tensor representing the distribution of stresses, momentum, and energy of whatever "things" there are about, material or otherwise.

Since (8.1) occupies so vital a position it is worthy of a general discussion, even at the risk of being here and there a little repetitive. First, its tensorial character, incidentally required by P1, automatically ensures its nonlinearity. That is just as well, for nonlinearity is the *sine qua non* without which no equation can imply equations of motion. A linear equation would admit as solution an arbitrarily linear superposition of any number of given solutions. One could then build up a solution representing various material distributions, permanently fixed with respect to each other; a conclusion at variance with what is observed. I am not saying, by the way, that nonlinearity is a sufficient condition for an equation to imply equations of motion. It so happens that (8.1) forces the vanishing of the divergence of T^{ij} simply because that of P^{ij} always vanishes as a consequence of the invariance of L under arbitrary coordinate transformations. To implications of this I shall return shortly in more detail. For the moment I merely repeat that it manifestly represents a restriction on the possible temporal behavior of whatever T^{ij} represents, a consequence which is a vital component of the whole conceptual scheme.

The content of *P*1 is not yet exhausted: it demands the absolute form invariance of T^{ij}. Evidently we previously left this requirement understood and we now make it explicit. Straight away, this allows T^{ij} to depend only on the field functions which are subject to dynamical equations, for no absolute[II] objects must be present; nor can it depend explicitly on the coordinates. That T^{ij} can always be constructed has to be taken for granted. Familiar special-relativistic field theories—recall that they are supposed to afford a local description of phenomena, referred to local inertial frames— contain a corresponding object. It often arises in the first place in the course of considerations concerning the local conservation of energy and momentum, possibly via the canonical energy tensor. Although it generally contains a skew-symmetric part, there are prescriptions for "removing" this. An analogous, though less straightforward, situation exists when T^{ij} represents material substances which are being described phenomenologically. When the substance is some fluid, for example, one also has "fields," namely, not only the density field, pressure field, viscosity fields, and the like; but, as well, thermodynamic fields such as the temperature and entropy fields. Of course, the presence of these thermodynamic quantities must give rise to serious reservations concerning the adequacy of any phenomenological description under conditions far from equilibrium. For purposes of orientation at least, one often contemplates so-called perfect fluids in whose description viscous effects are then ignored, as is the contribution of the heat flux to T^{ij}, which so reduces to the form

$$T^{ij} = (\rho + p)u^i u^j - pg^{ij}. \tag{8.2}$$

ρ is the invariant energy density, p the hydrostatic pressure, and u^i the local velocity of the fluid. At any rate, whatever the particular case in hand may be, we now follow the precedent set in the Maxwellian case, the prescription being this: transcribe the T^{ij} from the appropriate local theory which pretends that spacetime is locally exactly Euclidean to arbitrary curvilinear coordinates; $\hat{g}_{ij}$, say, being the transform of η_{ij}. Then interpret $\hat{g}_{ij}$ consistently as g_{ij} and identify the resulting tensor with that on the right of (8.1).

This step looks like an almost inevitable consequence of what went before, including appeals to $P3$, and it seems straightforward enough; but there are possible snags. Not only T^{ij}, but, to maintain consistency, the whole theory of which it is part, has to be transcribed; and when $\hat{g}_{ij}$ becomes g_{ij} ambiguities involving covariant derivatives may arise. For example, the transform of an expression such as $w_{i,jk}$ may be written indifferently as $w_{i;jk}$ or $w_{i;kj}$; but after the transition from $\hat{g}_{ij}$ to g_{ij} these two alternatives are no longer equivalent. Worse, the mere uncritical replacement of $\hat{g}_{ij}$ by g_{ij} may lead to inconsistencies. In place of the second of the equations $f_{ij} = 2A_{[j,i]}$, $\eta^{jk} f_{ij,k} = 0$ one may adopt the wave equation $\eta^{mn} f_{ij,mn} =: \Box f_{ij} = 0$, but after transcription to arbitrary coordinates and passage from $\hat{g}_{ij}$ to g_{ij}, one has three mutually inconsistent equations. The cause of the trouble is that the first two equations imply that $\Box f_{ij} := g^{kl} f_{ij;kl}$ must be equated to an expression which contains the Riemann tensor linearly. Confronted with this state of affairs one might simply insist that the rules of transcription are in the first instance to be applied to expressions involving at most first covariant derivatives, all second order equations being regarded as "derived." Even if this directive leaves sufficient elbowroom, there are cases where one is still confronted with virtually unsurmountable internal inconsistencies. These apart, one may be able to help oneself by appealing to $P3$ in the form of the principle of minimal coupling to resolve ambiguities. This principle has to be appealed to in any event to exclude the presence of terms additional to T^{ij} already defined; otherwise, on what grounds is one to reject an expression such as RT^{ij}?

It seems that as far as the right hand member of (8.1) is concerned I have ignored $P2$. The reasons for this are as follows. In the first place, whereas we had strong reasons for requiring P^{ij} to be a functional derivative there is no urgent motivation for insisting on the existence of a matter Lagrangian M which might generate T^{ij}. To be able to replace (8.1) by a variational principle might make one feel satisfied by the degree of formal elegance so attained. Yet, nothing very tangible would be gained. It seems to be sometimes taken for granted that one can always find a function M subject to $P1$ such that $T^{ij} = 2\delta M / \delta g_{ij}$; but whether such an assumption is justified is a moot point. At any rate, I propose to disregard this question, being satisfied with the direct availability of T^{ij}.

To the extent that $P2$ is being ignored here one might hold that one is also ignoring $P3$ if one takes the view that a theory wholly subject to $P2$ is simpler than one which is not. In parentheses, one also says at times that it "has greater aesthetic appeal" or "is more beautiful." Such phrases, however, surely reflect no more than loose, subjective, qualitative measures of simplicity, and we are back to *P*3. It is simply a fact of life that there is vagueness inherent in the interpretation of heuristic regulative principles.

With respect to *P*1 the position is much more clear-cut. Still, even here one may have some lingering doubts. What I have in mind in saying this concerns the presence of constants, that is, of numerical factors which, though universal, have values which are ultimately determinable only by experiment. The constant λ in (8.1) is an excellent example of this. $\lambda^{-1/2}$ is a time lapse—has the dimensions of time, if you like—but about its numerical value we can say nothing as yet. Should one however reject λ after all on the grounds that it represents the introduction of an absolute[IV]—and trivially absolute[II]—element into the theory, in contravention of *P*1? I think not, for if we do so we must likewise reject any "absolute constants" from T^{ij}. In the theory as it stands T^{ij} will, however, inevitably contain such constants, constants of a phenomenological character, like Boltzmann's constant, and, more seriously, constants such as the electronic charge. How can we throw these away? There are only two possibilities. The first is this: one rejects the need for T^{ij} altogether by adopting some extreme monistic ontology which holds that spacetime is the only existent in the world; a kind of supersubstantivalist position according to which everything else which we regard as existing is merely a derivative of this—granted that the phrase "we regard" is not a *non sequitur* here. The beginnings of a corresponding program—geometrodynamics—have been attempted, only to be rejected; not surprisingly, if only because the structure of the world as it actually appears to be is too complex in all its aspects to be describable on the basis of such an oversimplified ontology. The second possibility is to replace the general theory here considered by a "unitary theory" with a formally analogous but extended structure which, resting on the existence of a basic object more complex than g_{ij}, obviates the need for any energy tensor to appear. Einstein's own efforts in this direction appear to have proved fruitless. Whether more recent, more ambitious, attempts will ultimately prove successful remains to be seen.

Let us take stock. The discussion of various points raised previously yet needs to be resumed, and there are other issues of principle that demand attention. To one or two of these we shall come almost immediately, but there are others. For example, it would be desirable to answer the following questions. Can one legitimately speak of T^{kl} as a "source" of the "field" g_{ij}? Can one say that T^{kl} is the "cause" of this field? When $T^{kl}=0$ should one say that g_{ij} is "coupled to itself" or that "it is its own source"? However, I think all this should be left till later, since any further general discussion would be made largely pointless if the putative theory we have developed turns out to be totally at variance with the results of observation. This might very well be so on the nonlocal level for although the formalized results of observation went into the construction of the energy tensor, all such results were merely local. The only nonlocal ingredient is the regulative principle *P*1 which reflects our empirical knowledge that local theories, appropriately formulated according to fixed prescriptions, are indifferent to the choice of local inertial frames. In short, the task now before us is to investigate how nonlocal observational results might be predicted and to compare such predictions with what observation actually shows to be the case.

Alas, it is to all intents and purposes impossible to do justice to it. It will be less cumbersome to explain what I mean if I simply refer to a set of functions g_{ij} and T^{kl} which jointly satisfy the equation of consistency (8.1) as a "*g-T* pair." Any such pair constitutes a "solution" of (8.1). Here a peculiar, almost trivial, possibility seems to be at hand; choose any "reasonable" set of functions g_{ij} and substitute these into (8.1). As a result one has the T^{kl} as explicit functions of the coordinates. It is true that now we have, in fact, a *g-T* pair but from a realistic point of view it is an illusion. How are the "things"—the physical system—to which T^{kl} belongs to be identified? It cannot be done: the generic form of T^{kl} corresponding to some system of interest must be specified first, but then one has no assurance at all that this will be compatible with the functions already obtained. If, for example, the system is an electromagnetic field then one must necessarily have $T^i_{\ i}=0$ and $T^i_{\ k}T^k_{\ j}=\frac{1}{4}\delta^i_{\ j}T^k_{\ l}T^l_{\ k}$, other conditions apart. Likewise, if one imagines the system to be an ideal fluid, why should the computed functions T^{ij} be compatible with (8.2), especially since one has to have $\rho>0$? Occasionally the view has been expressed that (8.1) may be looked upon as providing a test for the presence of energy and momentum: a region of spacetime if void if and only if P^{ij} vanishes within it. That is all very well, but when $P^{ij}\neq0$ *what* is there to be found? What is the nature of the physical system which is supposed to be there? There is no answer—it cannot be recovered from a mere knowledge of the functions $T^{ij}(x^k)$.

In short, the generic form of T^{ij}, appropriate to whatever system happens to be of interest, must be prescribed at the outset. Further it must be supplemented by those constitutive equations which may be required to make (8.1) into a closed set of differential equations and this set must then be solved, subject to selected boundary, initial, and continuity conditions. Leaving the ticklish question of boundary conditions aside for the time being, one has a problem which is entirely intractable in practice. Let me explain what I am trying to say here. One can certainly find some solutions which have $T^{ij}=0$. The physical interpretation of any of these is carried out by investigating the motion of test objects, that is to say, of free particles and light pulses. One must not let the use of the term "particle" blind one to the fact that one has after all a material object of some kind and the maintenance of inner consistency demands that it should itself be accounted for by the energy tensor, and therefore by g_{ij}; but it is not. When $T^{ij}\neq0$ the situation is hardly more favorable. Certainly no exact solution is known which represents just two separated bodies.

We are faced with the inevitable conclusion that one cannot manage without approximative procedures, except possibly at the stage of finding a g-T pair by itself, that is, before contemplating more general solutions which would exactly represent also "test objects." To begin with, then, let us consider a test particle P—"test particle" rather than "free particle" merely to emphasize the present context. Consistently with the notion of a particle contemplated at the beginning of the third lecture, P is to be thought of as a free, nonrotating material distribution which is sufficiently small in size and has a sufficiently small inertial mass. What is all this intended to mean here? By way of explanation, let the time lapse corresponding to a light pulse traveling to and fro between its extremes be τ'. The inertial mass m is defined locally in the usual way, say by experiments involving mutual elastic collisions of various particles. Further, let g_{ij}, T^{kl} be a given solution of (8.1) in which no test particle is represented, and let $\tilde{g}_{ij}:=g_{ij}+h_{ij}, \tilde{T}^{kl}:= T^{kl}+\delta T^{kl}$ be another solution which can be interpreted as representing a system closely resembling the first, but also including a test particle. Then τ' and m must be so small that in any expression involving terms which are linear and terms which are nonlinear in the h_{ij} and their derivatives the linear terms completely dominate all others, so that the effects of the latter are negligible. Under these circumstances one has a linear differential equation for h_{ij}, the explicit form of which—by no means simple—need not concern us here. On the other hand, one must have $\tilde{T}^{ij}_{\;:j}=0$, where the use of the colon serves as a reminder that the covariant derivative refers to $\tilde{g}_{kl}$. To the required order, therefore, $(\delta T^{ij})_{;j}+T^{jk}\delta\Gamma^{i}_{\;kj}+T^{ik}\delta\Gamma^{j}_{\;kj}=0$. The test body in vacuo—a region in which T^{ij} vanishes—is therefore subject to the condition $(\delta T^{ij})_{;j}=0$.

What might one now take as a possible, sufficiently detailed, physical model of a test particle? A small piece of a solid of some kind? That would be difficult to describe phenomenologically. A fluid drop? We have the same problem. The motion of free drops is complex and is likely to introduce irrelevancies. Then why not a system free of internal forces, such as a small pressure-free cloud of dust particles? Here I am running into circularity. I cannot describe these particles, or if I could, any of them might have served as a test object in the first place. The way out of this impasse must inevitably be imprecise. We return to the fluid drop, but simplify matters by taking the fluid to be not only perfect but pressure free as well. All mutual forces between elements of the fluid are thus effectively absent. To understand this state of affairs better, contemplate the local description of a part of the fluid, temporarily supposing it to be electrically charged. To be consistent, the charge density has to be so small that the electrostatic interaction between the elements of the fluid can be entirely disregarded. Now let the fluid be placed into an intense, regular electrostatic field. Its individual elements then react separately to the external electrostatic force. Nevertheless, the fluid distribution remains continuous and a fluid drop remains a fluid drop though its shape and volume will change, the density ρ being allowed to take care of itself. However, for a fluid drop to qualify as a test object, a supplementary limitation has to be imposed, as follows. Since the drop—that is, the value of τ'—has already been taken to be sufficiently small, it can always be referred to a local inertial frame $\mathfrak{I}$ and it makes sense to require all of the fluid elements to be initially fixed in $\mathfrak{I}$. Only a certain limited part of the subsequent history of the drop will be considered.

In view of (8.2) we now have $\delta \tilde{T}^{ij} = \rho u^i u^j$, with $\rho \neq 0$ only within some sufficiently narrow world tube. Then $(\rho u^i u^j)_{;j}$ must vanish, or

$$u^i(\rho u^i)_{;j} + \rho u^j u^i_{;j} = 0.$$

Transvect this throughout with u_i. Then, bearing in mind that $u_i u^i = 1$ and therefore $u_i u^i_{;j} = 0$, it follows that $(\rho u^j)_{;j} = 0$, and in turn that $u^j u^i_{;j} \equiv u^j u^i_{,j} + \Gamma^i_{jk} u^j u^k = 0$. Under the present circumstances we may, for descriptive purposes, represent the motion of the drop sufficiently closely by the motion of one of its "infinitesimal elements." If x^i and $x^i + dx^i$ are two neighboring events in its history and ds the interval between these, $u^i = dx^i/ds$ and $u^j u^i_{,j} = du^i/ds$. Thus x^i obeys the equation

$$\frac{d^2 x^i}{ds^2} + \Gamma^i_{jk} \frac{dx^j}{ds} \frac{dx^k}{ds} = 0, \tag{8.3}$$

which is the equation for a geodesic of the "background metric" g_{ij}. This is precisely the equation which previously turned up by simply omitting the subscript 0 throughout the equation (5.5). Furthermore, we shall have consistency with prescriptions made long ago if we insist that the particles constituting a local inertial frame be themselves the kind of test bodies which we have just described in such detail.

We are now in the happy position of not having had to invoke the so-called geodesic principle by fiat. Let me repeat that the introduction of such a principle would run counter to the conceptual integrity of the theory as a whole. This I have stressed repeatedly. Moreover, we have only managed to obtain (8.3) under very restrictive circumstances. That is as it should be. If internal forces are admitted, a particle may rotate about itself and there is no obvious reason why its history should be a geodesic: in fact in general it is not. Again, if the mass of a test particle is finite, the idea of a background metric loses its meaning, which is as much as to say that one cannot have a test object whose mass is "finite." In short, rather than regard with disfavor the multitude of assumptions to which our derivation of (8.3) was subject, it is the geodesic principle which we should reject.

These remarks cast a shadow over the widespread use of delta functions in the representation of particles. That this can be no more than a purely formal device is obvious. Moreover, world-lines of particles are then singular, and on the face of it the idea of a background metric no longer makes sense. Evidently one has to investigate carefully how meaningful results may nevertheless be obtained in these circumstances. In fact, an extensive literature exists surrounding the question of the equations of motion of "finite mass particles," represented by appropriately defined delta functions. It involves calculations of enormous detail. I shall not concern myself with these here, if only because as a matter of principle I prefer those approximative schemes which confine themselves entirely to continuous material distributions whose actual realistic internal constitution is taken into account.

Such methods rely on the integration of (8.1) by a recursive procedure. The generic form of the energy tensor is of course prescribed and such constitutive relations, conservation laws and the like which are required to supplement (8.1) so as to make the whole into a closed set of equations, are explicitly set down. Next, all field functions are written as generic series in ascending powers of a suitable "small" numerical parameter ϵ. For example

$$g_{ij} = g^{(0)}_{ij} + \epsilon g^{(1)}_{ij} + \epsilon^2 g^{(2)}_{ij} + \cdots . \tag{8.4}$$

Note that g_{ij} and $g^{(0)}_{ij}$ here are the same as the $\tilde{g}_{ij}$ and g_{ij} which occurred earlier in this lecture, and all other terms jointly make up h_{ij}. Thus all terms on the right of (8.4) other than $\epsilon g^{(1)}_{ij}$ were neglected previously. Proceeding in this way, (8.1) becomes

$$\sum_{n=0}^{\infty} (P^{(n)}_{ij} - \kappa T^{(n)}_{ij}) \epsilon^n = 0.$$

The factor multiplying $\epsilon^n (n = 0, 1, 2, \ldots)$ is then required to vanish individually, so that (8.1) splits up into the sequence of equations

$$P^{(n)}{}_{ij} - \kappa T^{(n)}{}_{ij} = 0 \qquad (n = 0, 1, 2, \dots). \tag{8.5}$$

Let it now be granted that one has a solution $g^{(0)}{}_{ij}$ of the zeroth order equation, $n = 0$, and therefore that all the zeroth order quantities which enter into $T^{(0)}{}_{ij}$ are known. Wherever they occur in $P^{(1)}{}_{ij}$ and $T^{(1)}{}_{ij}$ they are now known functions, and one can go on to solve the first order equation $P^{(1)}{}_{ij} = \kappa T^{(1)}{}_{ij}$. In principle this process can be continued indefinitely, but one has to presuppose that convergence of the various series obtains. This is of course a weak point of any such procedure, for to prove convergence is a very knotty problem indeed. At any stage, the equations arising from $T^{ij}{}_{;j} = 0$ must be satisfied. Explicitly, the factor multiplying ϵ^n in the expression

$$\sum_{p,q} g^{(p)jk}(T^{(q)}{}_{ij,k} - 2\sum_r \Gamma^{(q)l}{}_{k(i} T^{(r)}{}_{j)l} \epsilon^r) \epsilon^{p+q}$$

must vanish for every $n = 0, 1, 2, \dots$. These conditions ultimately imply the equations of motion, correct to a certain order, depending on the number of recursive steps explicitly included in the calculations. How this goes is illustrated by our previous investigations of the small incoherent fluid drop, although this is unfortunately a rather trivial case. To go substantially beyond this would, however, land us in a forest of formal detail. From the start it was our avowed intention to keep the compass of formal calculations to a minimum in these lectures, and I therefore confine myself to one or two incidental remarks.

The calculations have a solution of the zeroth order equation as their starting point. Here we seem to be in trouble. In the context of the motion of a test object it sufficed to say "let $g^{(0)}{}_{ij}$, $T^{(0)kl}$ satisfy (8.1)"—it wasn't necessary to know what these functions actually looked like. Now, however, we are confronted with the realistic situation in which the presence of everything must be allowed for from the start. But was not the whole object of the recursive procedure to find such a solution? Indeed, it was and we are confronted with circularity. To break the circle, some further condition has to be imposed. To this end, let us reflect on the form of R_{ij}. From what we already know—recall equation (7.21)—

$$R_{ij} = g^{kl}(g_{i[j,k]l} + g_{l[k,j]i}) + N_{ij}, \tag{8.6}$$

where N_{ij} is an expression quadratic in the first derivatives of the g_{mn}. In principle one can always write (8.6) as

$$R_{ij} = \eta^{kl}\left(h_{i[j,k]l} + h_{l[k,j]i}\right) + N'_{ij},$$

where $h_{mn} := g_{mn} - \eta_{mn}$ and N'_{ij} represents all terms which do not depend linearly on the h_{mn} and their derivatives. Now, when the physical circumstances are such that these are sufficiently small in absolute value, N'_{ij} may simply be neglected and then R_{ij} reduces to an expression which is merely a sum of second derivatives of the h_{mn}. To simplify the resulting expression for g_{ij}, introduce the following abbreviations:

$$l_{ij} := h_{ij} - \tfrac{1}{2}\eta_{ij}h, \qquad l_i := l_{i,k}^{k}, \qquad h := \eta^{ij}h_{ij}.$$

Indices are here juggled with η_{mn}, of course. It then emerges that

$$-2G_{ij} = \Box l_{ij} - 2l_{(i,j)} + \eta_{ij}l^{k}_{,k}.$$

We are already aware that in the process of solving (8.1) coordinate conditions of some kind have to be imposed. We may as well do so now. By inspection, the condition $l_i = 0$ is a convenient choice, for the G_{ij} then reduce simply to $-\tfrac{1}{2}\Box l_{ij}$.

Now, if, as assumed, h_{ij} is sufficiently small, consistency requires that T_{ij} be sufficiently small. In particular it may even be zero in some regions; but then $\lambda\eta_{ij}$, that is to say λ, must also be sufficiently small. It seems advisable to save ourselves a lot of formal difficulties by taking its value to be zero for the time being. In due course we shall resurrect it, if only to confirm that we have not committed a grievous error in the present context. The equation to be solved is now

$$\Box l_{ij} = -2\kappa T_{ij}. \tag{8.7}$$

LECTURE 9

At the end of the preceding lecture we arrived at the equation (8.7) to which the general equation (8.1) reduces in consequence of (1) writing $g_{ij} = \eta_{ij} + h_{ij}$ and assuming the physical conditions to be such that all terms which depend nonlinearly on the h_{ij} and their derivatives may be rejected to a sufficient approximation; (2) setting $\lambda = 0$; (3) imposing the coordinate condition $l_i = 0$. I mention in passing that l_i may equivalently be removed by means of a suitable infinitesimal coordinate transformation, a process analogous to the removal of unwanted terms in Maxwell's theory by means of gauge transformations. Each of the component equations of (8.7) has the familiar form of an inhomogeneous wave equation. It has the retarded solution

$$l_{ij} = -(\kappa/2\pi)\int R^{-1}T_{ij}(\bar{x}^1, \bar{x}^2, \bar{x}^3, \bar{x}^4 - R)d\bar{x}^1 d\bar{x}^2 d\bar{x}^3, \qquad (9.1)$$

where $R^2 = \delta_{ab}(x^a - \bar{x}^a)(x^b - \bar{x}^b)$. Two points need to be borne in mind here. First, the preference given to the retarded solution is not mandatory and it rests on some possibly unreliable notions concerning causality; second, on the right of (9.1) one must, strictly speaking, add the general solution $\hat{l}_{ij}$ of the homogeneous equation $\Box \hat{l}_{ij} = 0$.

Now, since (8.7) is linear, distinct solutions may simply be superposed to yield new solutions. As discussed long ago, (8.7) can therefore not imply any sensible equations of motion. Consistently with this, on forming the divergence of both members of (8.7) the vanishing of l_i implies that $T^j_{i,j} = 0$. If one now pursues an argument exactly analogous to that leading to equation (8.3), the motion of a pressure-free "test drop" will be subject to the equation $d^2 x^i/ds^2 = 0$: it never "knows" anything about any other matter which may be about. This conclusion is contrary to the facts. The contradiction is however only apparent, to the extent that "knowledge" of this other matter is contained in terms which are nonlinear in the h_{ij} and their

70

derivatives; and these we chose to ignore. The hiatus may be overcome by using equation (8.3), taking the background metric in this case to be $g_{ij} = \eta_{ij} + l_{ij} - \frac{1}{2}\eta_{ij}\eta^{mn}l_{mn}$, with the l_{ij} given by (9.1). In this way one is implicitly taking previously rejected nonlinear terms into account.

Even at the level of the lowest order approximation one is confronted, of course, with the need to formulate suitable boundary conditions. How else is one to extract from $\hat{l}_{ij}$ that solution of the homogeneous equation which is relevant to the particular physical circumstances in hand? This problem arises anew at every step of the iterative process and is the cause of much perplexity. I have already mentioned that no exact two-body solution is known. The two-body problem is so important that one will therefore try to solve it by iterative approximation, but it is difficult to go far in this direction. In the first place, because of the presence of formally wavelike terms arising from $\hat{l}_{ij}$, physical intuition—based on our knowledge of comparable electromagnetic situations—would suggest the possible existence of some kind of radiation reaction. Perhaps two like bodies which would otherwise revolve about each other at a fixed distance apart might on this basis be expected gradually to approach each other. But will they, and, if so, at what rate? Having obtained some approximate solution containing "radiative terms" how can one say with certainty that these "belong" to the system in hand and do not contain contributions which have originated elsewhere?

In these remarks I have deliberately avoided any definite, detailed physical commitment. Nowhere have I spoken of "energy," for instance. The energy of what? Of the field perhaps? Field?—we have no field in the sense in which one has a Maxwell field. Whenever I have used the term "field" in the context of g_{ij} I have done so as a matter of mere verbal convenience. This remark is to be understood as follows. The classical fields—the electrostatic field, for example—in the first instance had, so to speak, a subjunctive existence. Let P be a particle satisfying all the criteria of being free except in as far as it carries an electric charge. Then *if* P were placed at some point it *would* be subject to a force depending on the value of the field intensity there. In particular, when the latter is zero there is no force. From all our previous discussions it should be abundantly clear that the g_{ij}-field is entirely different. Whatever the values of the g_{ij} and their derivatives, when there is no electric field, P never experiences a force. At the risk of endlessly repeating myself: the "true" fields subjunctively quantify the extent to which a given particle P is not free, granted that P would be free if any charges it may carry were neutralized. They are responsible for the accelerations of P but the idea of this acceleration is meaningless unless one can recognize unaccelerated, that is, standard motion; and it is the g_{ij}-field which provides the standard history of P. Concomitantly, this field, unlike the "true" fields, cannot be absent, cannot be zero—it would simply be nonsensical to contemplate a region in which $g_{ij} = 0$. Sometimes the g_{ij}-field is spoken of as "Führungfeld"—guiding field—but even that seems a bit indigestible with its forcelike overtones, for it conjures up a vague picture of a particle moving otherwise than it does were it not for the "action" of the leading field. But what is this other motion? We cannot say, it does not make sense except in as far as one might *arbitrarily* specify the history of a free particle by *prescription*; but that is where we came in and I won't rehearse this possibility again.

Back to energy. In Maxwell's theory there appears a field energy tensor $T_M{}^{ij}$. It arises from considerations involving the rate of change of energy-momentum of a set of charged particles produced by the forces exerted by the field. If overall energy-momentum conservation is required, one has to ascribe stresses, momentum, and energy to the field itself; and their densities collectively constitute $T_M{}^{ij}$. In a sense the field, initially subjunctive, is superficially hypostatized. At first sight there is no corresponding argument which might allow one to construct an object t^{ij} occupying a position vis-à-vis the g_{ij}-field of the kind which $T_M{}^{ij}$ occupies in Maxwell's theory. For, once again, the g_{ij}-field exerts no force and we cannot even begin to argue as before. Even from a purely formal point of view we are in immediate difficulties. t^{ij} must be a function of the g_{ij} and their derivatives, so that if one requires it to be a tensor it must be a concomitant of the Riemann tensor and possibly its covariant derivatives. Such an object would hardly qualify as a straightforward counterpart to the Maxwellian $T_M{}^{ij}$. In particular, Bel's so-called superenergy tensor is a quadratic function of the components of the Riemann tensor and when $R_{ij} = 0$ has certain algebraic and differential properties closely analogous to those possessed by $T_M{}^{ij}$. Yet it does not qualify, for it has valence four, not two. Perhaps, then, one should drop the demand that t^{ij} be a tensor. This seems out of keeping with the ideas underlying the theory as a whole, to say the least. I won't pursue this now but return to it as occasion arises. Enough has been said to justify my reluctance to talk about the possibility of radiation of "energy" from a two-body system.

The fact is that so far no foolproof, entirely unobjectionable approximative scheme has been devised to solve the two-body or more complex problems. Some calculations show that the bodies gradually approach each other—we shall derive one such result in the fifteenth lecture—while others would have it that they recede from each other. The rates of approach vary between different methods—in short the theoretical situation is inconclusive. The difficulties surrounding the imposition of boundary conditions and those engendered by the occasional appearance of divergent integrals have yet to be overcome. Furthermore, various mathematical inconsistencies have crept into all manner of procedures among which one distinguishes between so-called fast motion and slow motion methods and between those which treat particles as singularities and those which do not. One cannot even be sure that an exact solution corresponding to any given approximate solution exists. Add to this unresolved questions of convergence and obscurities regarding the equivalence of formally different results corresponding to the imposition of alternative coordinate conditions, and you will be convinced that much work needs to be done before a clear and conclusive picture can emerge.

These incidental remarks out of the way, it remains to consider, however briefly, the question of test light pulses within the theory. Recall that test particles and test light pulses must, strictly speaking, be accounted for by the T^{ij} on the right of (8.1). We disposed of the test particle—the small, pressure-free fluid drop—in the last lecture. For a light pulse, then, δT^{ij} will be the Maxwell tensor $T_M{}^{ij}$ under the assumption that the electromagnetic field is sufficiently weak. Since there are no electric charges $f^{ij}{}_{;j} = 0$ and this entails $(\delta T^{ij})_{,j} = 0$—recall the seventh lecture. What we have to do, therefore, is to solve Maxwell's equations in the geometrical optical limit, the metric g_{ij} contained in these equations being fixed, of course. The equations are with advantage taken in the form

$$f_{[ij,k]} = 0, \qquad f^{ij}{}_{;j} = 0. \tag{9.2}$$

Then, as usual without insisting on mathematical rigor, one seeks a solution of these equations which represents a locally approximately plane wave of sufficiently high frequency. To this end write $f_{kl} = a_{kl} e^{i\eta S}$, the physical field being understood to be the real part of this. Upon letting the parameter η become sufficiently large the spacetime variation of the amplitudes a_{kl} will be very small compared with that of the real scalar phase ηS. In the limit $\eta \rightarrow \infty$ the equations (9.2) then require that

$$a_{[ij}s_{k]} = 0, \qquad a_{ij}s^j = 0, \tag{9.3}$$

where $s_i := S_{,i}$. Transvecting the first of these with s^k and taking the second into account, it follows at once that $a_{ij}s_k s^k$ vanishes. Granted that a_{ij} vanishes at most on hypersurfaces, s_k is therefore a null vector:

$$g^{ij}S_{,i}S_{,j} = 0;$$

and this is the eikonal equation of geometrical optics. Differentiating $s_i s^i$ covariantly, one further has $s^i s_{i;j} = s^i s_{j;i} = 0$. Light rays are the orthogonal trajectories of the family of null hypersurfaces $S = $ constant. If $x^i = x^i(w)$ is the parametric equation of a ray, $s^i = dx^i/dw$ and the equation $s^i s^j_{;i} = 0$ just obtained becomes

$$\frac{d^2 x^j}{dw^2} + \Gamma^j_{ik} \frac{dx^i}{dw} \frac{dx^k}{dw} = 0,$$

so that the rays are null geodesics. Here we must be a little careful: before we can conclude that a test pulse will travel along a null geodesic we have still to show that energy is in fact so propagated. This is not a trivial point in as far as ray direction and direction of energy propagation might be distinct just as they are, in general, distinct in an anisotropic dielectric. The question is resolved easily enough. Transvection with a^{ij} of the first of (9.3) shows that the bivector a_{ij} is null. On the other hand transvection with any vector t^k not orthogonal to s_i shows that a_{ij} is a simple bivector, that is to say, $a_{ij} = 2s_{[i} q_{j]}$, where $q_j := (s_l t^l)^{-1} t^k a_{kj}$. The real field tensor is therefore, in the present approximation, $f_{ij} = \mathrm{Re}(2s_{[i} q_{j]} e^{i\eta S})$. Then $T_M{}^{ij}$ becomes a scalar multiple of $s^i s^j$ which shows that energy indeed propagates along rays. In other words, the history of an element of a test light pulse is, on the level of geometrical optics, a null geodesic. Consistency with the considerations which first set us on our way is therefore safeguarded.

It is time to take stock of our position. Step by step a coherent theory about spatiotemporal relationships—the general theory of relativity—has emerged. That, however, is saying too much, for, as I already stressed halfway through the last lecture, it might be an empty shell: it might not correctly describe what is in fact observed to be the case. It is all very well to say that everything went through in so "natural" a way that one could not countenance such a possibility—this tended to be Einstein's own attitude—but in the end it won't do. Whereas the history of science seems to suggest that "artificial" theories, theories containing a wealth of ad hoc assumptions and which give the appearance of flagrantly violating P3, inevitably turn out to be grossly defective, the absence of such features unfortunately does not guarantee success. In short, the time has certainly come to examine what the theory before us actually predicts.

Having spent much time on the uncertainties surrounding approximative methods it seems hardly apposite to begin with anything other than one or two exact solutions of (8.1); exact, that is to say, except in as far as they do not strictly accommodate realistic test bodies. We do not expect to be able to obtain them except under highly specialized circumstances and this is true even for regions in which $T^{ij} = 0$. Many ingenious schemes for finding exact solutions have been devised, yet, as far as physically realistic situations are concerned, they have not met with much success. It is easy to see why this should be so. One is accustomed to the idea that differential equations governing some system are soluble exactly only when the system has certain symmetries. The Hamilton-Jacobi equation is a good example. When certain symmetries are present the choice of a suitably adapted coordinate system may lead to separability of the equation, whereas in the absence of symmetries one is stuck with an intractable problem. Unfortunately realistic systems rarely have symmetries one would like them to have. Nevertheless, to make progress possible, we shall be well advised to require our system S to exhibit the kind of symmetry which is most likely to lead to success. Surely one's first choice will be spatial spherical symmetry, for we know from experience that then, and sometimes only then, can field equations be solved explicitly—think of the electrodynamics of Born and Infeld for instance. What about T^{ij}? In the first place we shall be unrealistic and take it to be zero. In electrostatics, too, one sometimes takes all charges to be absent. Then one finds that the electrostatic potential $\phi = A + q/r$—in a customary notation—A and q being constants of integration. There is a singularity at the origin which is interpreted as representing a point charge. This is not really acceptable since charges were from the outset supposed to be absent and one ought to set $q = 0$. Apart from that, singularities indicate the breakdown of a theory. We must bear these remarks in mind since we shall have occasion to refer to them later.

It is natural to adapt the coordinate system to the condition of spherical symmetry which is supposed to obtain. In effect one is also imposing coordinate conditions. Formally spherical symmetry is here understood to mean that every event corresponds to an ordered pair of points, one on a unit 2-sphere S_2 and one on a 2-space U_2. In terms of the usual polar coordinates θ, ϕ the metric of S_2 is $d\omega^2 := d\theta^2 + \sin^2\theta d\phi^2$. If that of U_2 is $d\sigma^2$ I take the spacetime metric ds^2 to be the sum of $d\sigma^2$ and a negative multiple of $d\omega^2$. The correct signature of ds^2 is assured provided $d\sigma^2$ is indefinite. Then coordinates u, v can be so chosen that $d\sigma^2 = 2f du\,dv$, the factor f—like the factor multiplying $d\omega^2$—being functions of u and v. In short, the metric of any spherically symmetric spacetime can always be taken in the generic form

$$ds^2 = 2f(u, v)du\,dv - r^2(u, v)d\omega^2. \tag{9.4}$$

It is worth remarking that there has been no reference to a center. Also, the sign of the first term on the right has no significance, for independently of it the signature of the metric has the correct value -2.

The immense simplification which spherical symmetry entails is clearly visible: one now needs to find only two functions of two variables in place of six functions of four variables. I say six, rather than ten, since by means of coordinate conditions four of the ten functions g_{ij} can always be prescribed, for example, $g_{a4}=0$, $g_{44}=1$. The adoption of the particular generic form (9.4) of the metric is of course not mandatory. For example, if one goes over from u and v to new variables R, ξ^4, say, (9.4) becomes generically

$$ds^2 = -A\,dR^2 - 2B\,dR\,d\xi^4 + C(d\xi^4)^2 - D\,d\omega^2 \qquad (9.5)$$

where A, B, C, D are functions of R and ξ^4, with the conditions $AC + B^2 > 0$, $D > 0$ satisfied. In turn one may make the transformation $\xi^1 := R\sin\theta\cos\phi$ $\xi^2 := R\sin\theta\sin\phi$, $\xi^3 := R\cos\theta$, so that $R^2 = \delta_{ab}\xi^a\xi^b$, $R\,dR = \delta_{ab}\xi^a d\xi^b$, $R^2 d\omega^2 = \delta_{ab}d\xi^a d\xi^b - dR^2$. Mere inspection reveals all three quantities to be invariant under the set of substitutions $\xi^a = L^a{}_{a'}\xi^{a'}$ which leaves the first of them invariant. In other words, ds^2 is manifestly form invariant under the familiar three-dimensional orthogonal group, ξ^4 being the untransformed coordinate. It may suffice to remark that spherical symmetry of a metric is sometimes taken to mean in the first place its invariance under three-dimensional rotations. Then, setting $h_{ij}=0$ in (7.5) with ξ^a appropriate to infinitesimal rotations and $\xi^4 = 0$, one has a set of differential equations which lead again to (9.5).

We now have to set about solving the equation $G_{ij}=0$. As a matter of convenience use the alternative notation $x^1 := u$, $x^2 := \theta$, $x^3 := \phi$, $x^4 := v$. The evaluation of the Christoffel symbols is straightforward. If subscripts 1 and 4 on f and r denote derivatives with respect to x^1 and x^4, that is, u and v, there are only twelve which do not vanish identically. They are

$$\Gamma^1{}_{11}=f_1/f, \qquad \Gamma^1{}_{22}=rr_4/f, \qquad \Gamma^1{}_{33}=rr_4\sin^2\theta/f, \qquad \Gamma^2{}_{33}=-\sin\theta\cos\theta,$$

$$\Gamma^2{}_{12}=r_1/r, \qquad \Gamma^3{}_{23}=\cot\theta, \qquad \Gamma^3{}_{13}=r_1/r,$$

the remaining five being obtained from these by the mutual interchange everywhere of the indices 1 and 4. Now, although the equations $G_{ij}=0$ are here equivalent to $R_{ij}=0$ I shall write down the components of the mixed Einstein tensor, as we shall need to know them later on. They are

$$G^1{}_1 = G^4{}_4 = 2(rr_{14}+r_1r_4)/r^3f + 1/r^2, \qquad (9.6)$$

$$G^2{}_2 = G^3{}_3 = (ff_{14}-f_1f_4)/f^3 + 2r_{14}/rf, \qquad (9.7)$$

$$G^4{}_1 = -2(fr_{11} - f_1 r_1)/rf^2, \tag{9.8}$$

$$G^1{}_4 = -2(fr_{44} - f_4 r_4)/rf^2, \tag{9.9}$$

the remaining components vanishing identically.

To begin with, the equations $G^4{}_1 = 0, G^1{}_4 = 0$ immediately imply that

$$f = 2B(v)r_1, \qquad f = 2A(u)r_4, \tag{9.10}$$

where A and B are arbitrary functions. With these results $G^1{}_1 = 0$ becomes alternatively

$$(rr_4 + Br)_1 = 0, \qquad (rr_1 + Ar)_4 = 0,$$

whence

$$r_1 = -A + F(u)/r, \qquad r_4 = -B + G(v)/r.$$

The integrability condition on these is $F(u)r_4 = G(v)r_1$, or, in view of (9.10), $F(u)/A(u) = G(v)/B(v)$. Since u and v are independent variables we must have $F = bA, G = bB$ where b is a constant. Accordingly r must satisfy the consistent differential equations

$$r_1 = -A(u)(1 - b/r), \qquad r_4 = -B(v)(1 - b/r), \tag{9.11}$$

and when it does so

$$f = -2AB(1 - b/r). \tag{9.12}$$

(9.11) and (9.12) together already satisfy the remaining equation $G^2{}_2 = 0$ so that the solution of the spherically symmetric problem merely requires the integration of (9.11).

When $b = 0$ we have

$$r_1 = -A, \qquad r_4 = -B, \qquad f = -2AB.$$

Write $\int A(u) du =: -\mu$ and $\int B(v) dv =: -\nu$. Then, virtually by inspection, $r = \mu + \nu$ and $f\, du\, dv = -2\, d\mu\, d\nu$ so that

$$ds^2 = -4\, d\mu\, d\nu - (\mu + \nu)^2 d\omega^2.$$

Now make the change to independent variables $\rho := \mu + \nu, \tau := \mu - \nu$; and then

$$ds^2 = -d\rho^2 - \rho^2 d\omega^2 + d\tau^2,$$

which is manifestly a Euclidean metric, ρ, θ, ϕ being spherical polar coordinates. This case is therefore trivial and we henceforth exclude it: $b \neq 0$. Then we may further economize by setting $b = 1$, for b may eventually be restored from dimensional considerations.

The functions $A(u)$ and $B(v)$ may be chosen in any convenient way, for if we make a coordinate transformation of the form $\bar{u} = \alpha(u), \bar{v} = \beta(v)$ and then drop bars, we, in effect, merely replace A and B by other functions of u and v respectively. Since

$$dr = r_1 du + r_4 dv = -(1 - 1/r)(A(u)du + B(v)dv),$$

the choice $A(u) = -1/u, B(v) = -1/v$ seems a happy one, for then one has

$$r\,dr/(r-1) = dw/w, \tag{9.13}$$

where $w := uv$. Integrating (9.13)

$$(r-1)e^r = w, \tag{9.14}$$

where a constant of integration has been absorbed in w. Also now

$$f = -2r^{-1}e^{-r}. \tag{9.15}$$

We thus have finally the Kruskal-Szekeres metric

$$ds^2 = -4r^{-1}e^{-r} du\,dv - r^2 d\omega^2, \tag{9.16}$$

where the dependence of the function r on u and v is given by (9.14). Evidently r runs from 0 to ∞ as w runs from -1 to ∞, whereas r runs from $-\infty$ to 0 as w runs from 0 to -1; so that r is not a single-valued function of w.

Inspection of (9.16) reveals at once that the *only* singularity occurs at $w = -1$, that is, $r = 0$. To make sure that this is no mere quirk of the coordinate system it suffices to calculate the invariant $R_{ijkl}R^{ijkl}$. A constant factor apart, this turns out to be r^{-6}, so that the singularity at $w = -1$ is genuine. Elsewhere everything is regular and the derivatives to all orders of f

and r with respect to w, that is, with respect to u and v exist, and this is true in particular at $w=0$.

Unfortunately (9.14) cannot be resolved for r in terms of known functions, but one may have recourse to power series. For instance, taking $r>0$, Lagrange's theorem yields near $w=0$ the result

$$r=1+ \sum_{n=1}^{\infty} (-1)^{n-1}n^{n-1}(w/e)^n/n!, \qquad (9.17)$$

and r^2 and f may be written in the form of similar explicit series. That on the right of (9.17) converges when $-1<w\le 1$, or $0<r\lesssim 1.2785$. For sufficiently large r one has from (9.14) $r=\ln w - \ln\ln w$ to within additive terms which go to zero as $r\to\infty$. In this limit the metric then turns out to be Euclidean. That is not at all surprising when we reflect that letting r go to infinity is equivalent to taking $b=0$ in (9.11) and (9.12).

Ignoring the "angular coordinates" θ and ϕ, the metric (9.16) covers that region K of a Cartesian representative u, v plane which lies between the two branches of the hyperbola H whose equation is $uv=-1$. K is conveniently thought of as made up of four pieces: $I(u\ge 0, v\ge 0), II(u\ge 0, -u^{-1}<v\le 0), III(u\le 0, v\le 0), IV(-v^{-1}<u\le 0, v\ge 0)$. The geodesics pictured in K have θ and ϕ constant—we call them radial geodesics—and it will suffice for the time being to consider them alone. The radial null geodesics are $u=$ constant, $v=$ constant. Any timelike geodesic or indeed any timelike world-line must have positive ds^2. Therefore $du\, dv$ must be negative and so either $du>0, dv<0$ or $du<0, dv>0$. It suffices arbitrarily to choose the first of these alternatives since the mutual interchange of the two possibilities merely corresponds to the mutual interchange of the pair I and II with the pair III and IV. Any timelike world-line thus lies between null lines $u=$ constant, $v=$ constant and the future direction of its tangent points toward the lower branch of H. Consistency demands that the future direction of the null line $u=$ constant be downward and that of $v=$ constant from left to right.

A light pulse represented by the line $u=$ constant reaches the singularity H and its history ends there. As regards any material particle, its world-line may or may not cross the u-axis. If it does so its history will come to an end when it reaches the singularity H. The crucial point is here this: the history of any object, whatever forces may be acting on it—say by means of a rocket motor attached to it—must inevitably encounter H, that is to say, come to an end there, once the object finds itself in region II. The traffic is strictly one way: light and material objects can cross into II from I but not from II into I.

This is all very unfamiliar and we are failing to make much contact with the observational knowledge we possess. That is perhaps not surprising since we are implicitly concerning ourselves with conditions very far from Euclidicity. We must surely direct our attention for the time being to the more nearly Euclidean region described by (9.16). This we shall do in the following lecture, toward the end of which I shall briefly return to region II.

LECTURE 10

The nearly Euclidean region corresponds, as we already know, to large values of w. Now the manipulations arising from (9.16) are undeniably awkward and we shall do well to make a transformation of coordinates which will bring (9.16) to a form better suited to the purpose in hand. To this end, define a new variable $t:=\ln(u/v)$ so that $du/u-dv/v=dt$. On the other hand, from (9.14), $du/u+dv/v=r\,dr/(r-1)$. Therefore

$$4\,du\,dv/w=-dt^2+r^2dr^2/(r-1)^2.$$

Inserting this in (9.16) and eliminating w by means of (9.14), the metric finally becomes

$$ds^2=-\gamma^{-1}dr^2-r^2d\omega^2+\gamma dt^2 \qquad (10.1)$$

with $\gamma:=1-b/r$, having restored the constant b. Here r and t are now coordinates in place of u and v. When $r\to\infty$ one immediately ends up with a metric which is manifestly Euclidean. Certainly (10.1)—the so-called Schwarzschild metric—leads to much simpler manipulations than does (9.16), but we have had, alas, a heavy price to pay for this simplification. The metric has a singularity at $r=b$, but since this corresponds to w being zero it must be artificial and without physical significance. Still, it need not worry us, being concerned at the moment only with values of $r>b$.

I note in passing that in (10.1) the g_{a4} are zero, while all metric components are independent of t. Now consider any metric $ds^2=g_{ij}dx^idx^j$ with x^4 supposed to be the timelike coordinate. That means that if (x^a, x^4) and (x^a, x^4+dx^4) are two neighboring events in the history of a clock, the time lapse measured by it is $(g_{44})^{1/2}dx^4$. Then if $A[x^i]$ and $B[x^i+dx^i]$ are any two neighboring events, the change dx^4 of x^4 corresponding to the emission of a light pulse from A and its return there—bear in mind the revearsal of its direction at B—is the difference between the two roots of the equation $ds^2 = 0$ for dx^4. The corresponding time lapse as registered by a clock at A is $(g_{44})^{1/2}dx^4$ and this is—by definition, you will recall—twice the distance dl between A and B. Explicitly,

$$dl^2 = (-g_{ab} + g_{a4}g_{b4}/g_{44})dx^a dx^b. \tag{10.2}$$

The distance is evidently independent of time if the g_{ij} are independent of x^4. When this condition is satisfied the metric is "stationary" and if, in addition, $g_{a4} = 0$ it is "static." In particular the metric (10.1) has turned out to be static, simply as a consequence of its required spherical symmetry alone, for it was not necessary to require asymptotic Euclidicity, for this also emerged automatically. As a matter of fact, a theorem known as Birkhoff's theorem states that a spherically symmetric solution of the vacuum field equation is necessarily static. Here we must be a little careful: in the simplified form just stated the theorem is intended to refer only to the region $r > b$, that is, $w > 0$, of spacetime. When $w < 0$ the choice of a new variable $t := \ln(-u/v)$ in (9.16) leads to (10.1) again, where, however, $r < b$ now. Therefore $g_{44} < 0$ and x^4 is no longer timelike, meaning $g_{44}(dx^4)^2 < 0$; but when taking the conditions $g_{44} > 0$, $g_{4a} = 0$ to define the static character of a metric it is implicitly taken for granted that x^4 is timelike.

As regards physical implications, contemplate a star S which is spherically symmetric and therefore nonrotating. Suppose that S is not emitting any radiation. Then any disturbance of S—no matter how cataclysmic—which leaves its spherical symmetry unaffected cannot become known to an observer outside S. All he can do is to investigate the motions of test objects in the vacuum region surrounding S; but they are determined by the *fixed* metric (10.1).

One can free Birkhoff's theorem from the condition $w > 0$ by reformulating it altogether. Without becoming too technical, an appropriate statement is this: any spherically symmetric solution of the equations $G_{ij} = 0$ is equivalent to a piece of the solution (9.16). Incidentally, the condition for a metric to be a stationary or static can also be stated in a coordinate-independent way, namely in terms of the existence of an appropriate Killing vector ξ^i, that is, a vector satisfying Killing's equation $\xi_{(i;j)} = 0$—recall equation (7.10). It is hardly worth our while to enlarge upon this here. It is better to say a word about an assumption inherent in the argument leading up to (10.2). The clock at A is in general not free—its history is not a geodesic. Nevertheless one takes it for granted that the finite time lapse which it registers between two events A and B on its world-line is $\int_A^B ds$, under these or any other circumstances. This assumption—sometimes called the chronometric hypothesis—is a curious one: try dropping a clock to the ground from the Eiffel Tower. This suggestion is not as silly as it looks at first sight, for all we are doing is to think about the effects of strong forces on a clock. Whether such effects can always be allowed for is a moot point. Even when the clock is free it is not clear what might be the case under extreme circumstances, for example, as one approaches the singularity $w = -1$ in (9.16). One suspects that the chronometric hypothesis will then cease to be adequate and it may well be that such extreme conditions will make nonsense of a lot of other physics anyway.

It seems that I have been talking more and more in a way which implicity seems to take it for granted that the theory is valid. That won't do and it is urgently necessary to go about the task of comparing predictions of the theory with observational knowledge. To this end we naturally begin with investigating the history of a free particle P, that is, we have to find the timelike geodesics of (10.1). When solving (8.3) let $i = 1, 2, 3, 4$ correspond to r, θ, ϕ, t in that order. Derivatives with respect to r and s shall be indicated by primes and dots respectively. For $i = 2$ the equation to be satisfied is

$$\ddot{\theta} + 2r^{-1}\dot{r}\dot{\theta} - \cos\theta\sin\theta\dot{\phi}^2 = 0.$$

If initially $\theta = \pi/2$ and $\theta' = 0$, then $\theta = \pi/2$ permanently and by an appropriate rotation this state of affairs can always be arranged to obtain. Then the equations for $i = 3$ and 4 are

$$\ddot{\phi} + 2r^{-1}\dot{r}\dot{\phi} = 0, \qquad \ddot{t} + \gamma^{-1}\gamma'\dot{r}\dot{t} = 0.$$

The first integrals of these are

$$r^2\dot{\phi} = \text{constant} =: h, \qquad \gamma\dot{t} = \text{constant} =: c. \tag{10.3}$$

In place of the equation for $i = 1$ one may straight away use equation (10.1):

$$-\gamma^{-1}\dot{r}^2 - r^2\dot{\phi}^2 + \gamma\dot{t}^2 = 1. \tag{10.4}$$

$\dot{t}$ and $\dot{\phi}$ may be eliminated by means of (10.3) and $\dot{r} = hr^{-2}dr/d\phi$. With the substitution $u = r^{-1}$ (10.4) now becomes

$$(du/d\phi)^2 + u^2 = (c^2 - 1)/h^2 + (b/h^2)u + bu^3.$$

Upon differentiating with respect to ϕ this finally gives the equation

$$d^2u/d\phi^2 + u = b/2h^2 + \tfrac{3}{2}bu^2. \tag{10.5}$$

First consider values of r so large that the second term on the right can be altogether disregarded. Then, writing $\alpha := b/2h^2$ the equation has the solution

$$u = \alpha(1 + e\cos\phi), \qquad (10.6)$$

where e is a constant of integration and the origin of ϕ has been chosen appropriately. Choose $e^2 < 1$ in order to ensure that P does not recede to infinity. Then according to (10.6) the orbit of P is an ellipse of eccentricity e with the point $r = 0$ as nominal focus; nominal, because this point is in fact excluded here, together with a region surrounding it. This, however, is of no account: we may simply picture (10.6) on a Cartesian representative plane C with r, ϕ as polar coordinates, and then $r = 0$ is a perfectly good focus of the ellipse. Alternatively we may look upon (10.1) as the vacuum solution which, at $r = r_0$, say, joins smoothly onto a solution of $G^{ij} = \kappa T^{ij}$, where T^{ij} represents a regular, static, spherically symmetric distribution of matter S within $r \leq r_0$. The point $r = 0$ is then regular. Further, if the matter density is sufficiently small, spacetime will be nearly Euclidean everywhere. We shall consider composite solutions in due course.

Now, what about the neglected term in (10.5)? We could have retained it from the start and, subject only to the condition $0 < bu < 1$, solved (10.5) in terms of elliptic functions. However, we are at present only interested in situations where u is so small that the effects of the term $\frac{3}{2}bu^2$ on the right of (10.5) merely induce a small correction to (10.6). It therefore suffices to replace u by $\alpha(1 + e\cos\phi)$ in the term in question, which so becomes $\frac{3}{4}b\alpha^2[(2 + e^2) + 4e\cos\phi + e^2\cos 2\phi]$. Of these terms only the second will produce a significant effect, namely cumulatively by resonance, its period coinciding with that of the unperturbed solution (10.6). It therefore suffices to add to this the particular integral $u_1 = \frac{1}{2}A\phi\sin\phi$ of the equation

$$d^2u_1/d\phi^2 + u_1 = A\cos\phi,$$

where $A := 3b\alpha^2 e$. Thus

$$u = \alpha(1 + e\cos\phi) + \tfrac{1}{2}A\phi\sin\phi$$

$$\approx \alpha(1 + e\cos(\phi - A\phi/2e\alpha))$$

as long as $A\phi/2e\alpha \ll 1$. It does not matter that we do not know the precise physical significance of u: at pericenter it is a maximum and we can understand the orbit as pictured in C. The maximum following upon that when $\phi = 0$ occurs when $\phi - A\phi/2e\alpha = 2\pi$, or, nearly enough, when $\phi = 2\pi +$

$\pi A/e\alpha$. The orbit is then no longer an ellipse. The pericenter, instead of being fixed, advances by an amount $\delta := \pi A/e\alpha = 6\pi h^2\alpha^2$ in each revolution, a revolution being reckoned from pericenter to pericenter. If r_+ and r_- are the extreme values of r and a is the semimajor axes of the ellipse, (10.6), with $\phi=0$ and $\phi=\pi$, shows that $a=\frac{1}{2}(r_+ + r_-)=[\alpha(1-e^2)]^{-1}$. Again, ignoring the precession of the pericenter, the first member of (10.3), that is, $r^2d\phi=h\,ds$, has the integral $\int_0^{2\pi}r^2d\phi=2\pi\alpha^{-2}(1-e^2)^{-3/2}=hT$, in which T is the period of revolution of P. α and h may now be eliminated from δ in favor of a and T to give

$$\delta=\frac{24\pi^3a^2}{T^2(1-e^2)}.\tag{10.7}$$

This equation and that of the orbit in lowest order of approximation,

$$r=\frac{a(1-e^2)}{1+e\cos\phi}\tag{10.8}$$

together provide a suitable characterisation of the orbit of P in the "asymptotic region" $0<bu\ll1$.

Now, the crucial question is this: do we in fact encounter large scale orbits of slowly moving particles of the kind described by (10.7) and (10.8)? We do. According to Kepler's first law the trajectory of a planet—regarded as a test particle, and with the presence of all other planets ignored—is in fact described by (10.8), with the sun at the focus of the ellipse. Moreover, in the asymptotic limit within the framework of which we are operating, Kepler's second law $r^2\dot\phi=$constant has also appeared, as has the third, according to which a^3T^{-2} is the same constant for all planets, its value being $b/8\pi^2$.

That Kepler's laws have emerged is already very remarkable. Still, we have been careless in one respect, namely we have blandly ignored the presence of other planets when considering the motion of one of them. If we are to do better we can no longer regard the planets as test particles. Then, however, we seem to be landed with all the doubts and complexities surrounding approximative schemes which previously came to our attention. In the lowest order of approximation these fortunately all lead to the same result; a result which can also be obtained by a heuristic trick, as we shall see shortly. Let us therefore simply accept that equations are available from which the perturbations of the orbit of P due to the presence of the other planets may be calculated. Only after these have been allowed for can one claim the validity of Kepler's laws in any case.

But are they valid? Is the orbit of P an ellipse? Not quite. It used to be regarded as a disconcerting fact of observational astronomy that the perihelion of Mercury precesses at the rate of about 43 seconds of arc per century after the effects of the other planets have been taken into account. Small though this residual precession may be, it exists. Could it be that (10.7) just accounts for it? The smaller the eccentricity of an orbit the more difficult it is to locate its perihelion. The planet Mercury has the most eccentric orbit but even here e is only about 0.206. Taking Kepler's third law into account and approximating $1-e^2$ by unity, $a\delta$ has a fixed value for the various planets. Again, therefore, Mercury represents the most favorable case. It has $T=7.6\times10^6$ seconds and $a=193$ seconds. These values together with that of e, when inserted into (10.7), give $\delta\approx43$ seconds of arc per century, in astonishing agreement with its observational value. I say "astonishing" because the agreement exists between it and a value calculated from an equation containing no adjustable parameters. Similar calculations can be carried out for other planets and asteroids and within the rather large empirical uncertainties agreement between observational and predicted values again obtains.

The state of affairs which has unfolded itself is so remarkable that—exaggeratedly perhaps—one is led to the firm conviction that the theory is in fact valid; the agreement between theory and observation can be no "accident." Only under extreme circumstances, that is, when spacetime can in no sense be regarded as being even approximately Euclidean should we perhaps reserve judgment. There is of course still the possibility that while the motion of particles is correctly described, that of light pulses is not. We therefore need to consider null geodesics. On the right of (10.4) there now appears 0 instead of 1 and proceeding as before we arrive at the equation

$$d^2u/d\phi^2 + u = \tfrac{3}{2}bu^2. \tag{10.9}$$

As a first approximation the solution of this is the straight line $u=R^{-1}\cos\phi$ in the representative plane C, R being the distance of closest approach to the origin. To a second approximation we thus need to solve the equation

$$d^2u/d\phi^2 + u = \tfrac{3}{2}bR^{-2}\cos^2\phi.$$

Thus

$$u=R^{-1}\cos\phi + \tfrac{1}{2}bR^{-2}\left(2-\cos^2\phi\right),$$

or if x, y are Cartesian coordinates in C,

$$x = R - \tfrac{1}{2}bR^{-1}(x^2 + 2y^2)(x^2 + y^2)^{-1/2}.$$

The asymptotes to the track correspond to y being much greater than x, so that for them

$$x = R \pm bR^{-1}y.$$

The small angle δ^* between these is given by

$$\delta^* = 2b/R. \tag{10.10}$$

Having noted previously that $b = 8\pi^2 a^3 T^{-2}$ we may calculate its value from the data for Mercury which are already to hand. It turns out to be 9.83×10^{-6} second. If light from a distant star just grazes the sun's limb, $R = 2.32$ seconds and $\delta^* \approx 1.75$ seconds. This value is confirmed by observation and here also the theory is successful. Incidentally, for Mercury $b/a \approx 5 \times 10^{-8}$, which shows how minutely g_{11} and g_{44} differ from unity.

Just now we considered the track of a light pulse but we may also inquire into the character of light emitted from the sun as observed on earth. The history of a monochromatic pulse is in general described by the Einstein-Maxwell equations (7.15, 7.20), but for a test pulse the g_{ij} are not affected by the presence of the electromagnetic field. In a static spacetime Maxwell's equations have monochromatic solutions of the kind $f_{kl} = \phi_{kl}\exp(2\pi i \nu x^4)$, where the ϕ_{kl} are independent of x^4. One cycle therefore corresponds to $\Delta x^4 = \nu^{-1}$, both on the sun and on the earth. Emitter and receiver are fixed on the sun and earth, respectively. Therefore Δx^4 corresponds to a time lapse $\Delta s = \sqrt{g_{44}}\Delta x^4$. The physical period of vibration, that is to say, the wavelength λ_s as measured on the sun is therefore ζ_s/ν, where $\zeta_s := (\sqrt{g_{44}})_{\text{sun}}$, whereas when the radiation is received on the earth its observed wavelength is ζ_e/ν, with $\zeta_e := (\sqrt{g_{44}})_{\text{earth}}$. The ratio of these is

$$\lambda_e/\lambda_s = \zeta_e/\zeta_s =: 1 + z > 1, \tag{10.11}$$

so that one has a redshift, measured by the parameter z. For the sun z is so small that its reliable determination is difficult, but for certain stars other than the sun the validity of (10.11) is also borne out by observation. I mention in passing that much confusion has surrounded the derivation of (10.11) in the past not only because of indiscriminate uses of the word "time" but also because of an ambiguity in the notation. It would in fact be better to write λ_{AB} for the wavelength of a spectral line, identified by the atomic transition responsible for it, emitted by an atom at A as observed at B, so that $z = \lambda_{se}/\lambda_{ss} - 1$. It is, of course, an underlying fundamental assumption that $\lambda_{ss} = \lambda_{ee}$. Indeed, its validity is part and parcel of the fundamental position which we have accorded to atomic clocks in the whole scheme. In the context of redshifts and elsewhere one quite often comes across unqualified statements about the "slowing down" of standard clocks. I think one does well to treat these with circumspection.

Before passing on we must momentarily resurrect the constant λ which we provisionally set equal to zero at the end of the eighth lecture. When $\lambda \neq 0$ the metric (10.1) satisfies the equation $G_{ij} = \lambda g_{ij}$, granted that now

$$\gamma := 1 - b/r - \tfrac{1}{3}\lambda r^2. \qquad (10.12)$$

The last term on the right causes the appearance of an additional term $-\tfrac{1}{3}\lambda/h^2 u^3$ on the right of (10.5). Since in the lowest order of approximation the equation must still be satisfied by (10.6) the dominant effect of the new term can only be an additional precession δ_λ. In this sort of order of magnitude calculation it will suffice to take $e \ll 1$ for all the planets. Then it turns out that

$$\delta_\lambda = (\lambda/4\pi)T^2,$$

or about $4 \times 10^{20}\lambda$ seconds of arc per century for Mercury. If we somewhat optimistically accept that a precession of 1 second would be observable we conclude that $|\lambda|$ must certainly be less than 3×10^{-21} second^{-2}, but it may well be very much smaller or zero. It cannot be substantially larger or else the predicted planetary orbits would not resemble those actually observed. λ is likely to play a role only on a cosmical scale and it is in fact known as the cosmical constant.

Despite the paucity of conclusive tests of the theory—bear in mind that we have perforce been able to consider so far only observational evidence relating to nearly Euclidean spacetimes—the agreement between predicted and observed results, especially of the precession of perihelia, is surely compelling. At any rate, I shall take it for granted that the theory as a whole is in fact valid to the extent that any theory which disregards the implications of quantum mechanics can be regarded as valid. It also breaks down where solutions of the field equations involve singularities. Just what constitutes a singularity in general and how it is to be characterized is a question not easily answered and this is not the place to elaborate the difficulties and subtleties involved in it. Still, one will certainly wish to speak of "a singularity of the metric at a spacetime point A"—as we already did in the ninth lecture—under circumstances such as the following: (1) a timelike geodesic terminates at A after a finite lapse of proper time; (2) an appropriate invariant of the Riemann tensor becomes infinite at A, granted that this infinity is reflected physically in the existence of infinite tidal effects. Under these circumstances it seems legitimate to take the view that at A intelligible physics comes to an end. On the other hand, we must certainly take (9.16) seriously for all values of $w > -1$. Then, for example, (10.9) shows that a light pulse may have a circular track, the equation of the circle being $r = \frac{3}{2}b$. As usual we must avoid thinking of r as a "radius." Even superficially it won't do, for a radius is the distance from the center to points on the circumference: but here, what "center"? Furthermore, distances have throughout been treated as indirect, derived quantities, names for times of travel of to and fro light signals. Here no such to and fro journey is possible, as we have seen. This is a good example in support of the contention that it might be best to avoid talking about distances altogether. By way of further amplifying this remark, one is in difficulties even when the metric is static. Take the metric (10.1) for example, and let a light pulse be emitted from $A[r_1, \theta, \phi]$ when $t = t_1$. It is reflected at $B[r_2, \theta, \phi]$ when $t = t_2$ and returns to A at $t = t_1'$. Set $ds = 0$ in (10.1), so that $dt = \gamma^{-1/2}dr$. Then $t_2 - t_1 = \int_{r_1}^{r_2} \gamma^{-1/2}dr$, having taken $r_2 > r_1$. Since the metric is static, $t_1' - t_2 = t_2 - t_1$. With the usual prescription in mind, we might then ascribe to A and B the mutual distance $D_1 := \frac{1}{2}\sqrt{\gamma}(r_1)(t_1' - t_1)$, that is to say, the total time lapse, registered at A, corresponding to the light journey from A to B and back to A:

$$D_1 = \sqrt{\gamma}(r_1) \int_{r_1}^{r_2} dr/\sqrt{\gamma}(r).$$

The trouble with this is that there does not appear to be any good reason for not taking the distance to be the time lapse D_2 registered at B, corresponding to a light journey from B to A and back to B:

$$D_2 = \sqrt{\gamma}(r_2) \int_{r_1}^{r_2} dr/\sqrt{\gamma}(r).$$

It sounds dangerously like having a distance relation which is not reflexive. Sometimes the "distance" is understood to be $D^* = \int_{r_1}^{r_2} dl$—recall equation (10.2)—

$$D^* = \int_{r_1}^{r_2} dr/\sqrt{\gamma}(r), \tag{10.13}$$

a quantity often referred to as the proper distance. In general the measures D_1, D_2, and D^* are quite different from each other. As a matter of fact, when $r_1 \to b$, $D_1 \to 0$, $D_2 \to \infty$ while D^* goes to a finite limit greater than zero. When the metric is not stationary the ambiguities are more severe—but all of them are artificial. They arise out of our mistaken conviction that there must be a unique physical quantity which is an image of the intuitive notion of distance with which we grew up. This is an illusion. Under given circumstances we may refer to distances, as long as we say clearly what we are talking about; but we can manage without it. In astronomy on a galactic or cosmic scale we merely need to achieve a coherent description of what is observed: parallaxes, apparent diameters, apparent luminosities, redshifts, and the like. More of this later.

Since we now accept the validity of the theory we must take the region $-1 < w \leq 0$ of (9.16) seriously. This means that we must visualize the possible existence of objects which behave as stars in as far as planets in their vicinity execute the usual motions, but which, as we have seen, can only absorb but not emit matter and radiation, the null surface $w = 0$, the so-called event horizon, being effectively a one-way barrier. One speaks of "black holes," an unhappy terminology to which the French usage of calling them "occluded stars" is to be preferred. Whether they actually exist in nature is not firmly established as yet. They well may, for it is not easy otherwise to come to terms with the evolutionary history of certain stars once their nuclear power supply is exhausted. Such stars will presumably undergo catastrophic collapse and all its matter will end up in the region $w < 0$. Once inside it it will inevitably age into the singularity $w = -1$: I say "age," because in (10.1) r is a timelike coordinate when $r < b$. In the thirteenth lecture an opportunity will present itself to say a little more about black holes, so that I need not pursue this topic now.

Instead, I shall return to a certain inconsistency to which I already drew attention during the preceding lecture. We have solved the equation $G^{ij} = \kappa T^{ij}$, setting $T^{ij} = 0$ and requiring spherical symmetry, to end up with a solution of which one could not really assert that $T^{ij} = 0$ everywhere on account of the occurrence of the singularity at $w = -1$. Viewed somewhat differently, if the vacuum solution joins smoothly at $r = r_0$ on to a solution representing the distribution of matter S mentioned earlier, then as the density of S goes to zero the spacetime becomes Euclidean. Formally, at any rate, one might therefore wish to invent a singular energy tensor associated with (9.16), in analogy with the singular electrostatic charge distribution $2\delta(r)/r^2$ which implies the singular electrostatic potential $\phi = 1/4\pi r$ of a unit charge.

Let me show briefly how this may be done. Since f and r are to be functions of w alone, (9.8) and (9.9) read

$$(r'/f)' = -\tfrac{1}{2} r v^{-2} G^4{}_1 = -\tfrac{1}{2} r u^{-2} G^1{}_4,$$

primes denoting derivatives with respect to w. Then

$$-2r'/f = \int rv^{-2}G^1_4 dw.$$

On the other hand, from (9.14) and (9.15) it follows that the left hand member of this has the fixed value 1. This is the case even when $b=0$, as (9.11) and (9.12) show. Consequently G^1_4 and G^4_1 are to be taken as zero. Next, (9.6) may be written

$$\tfrac{1}{2}r^2 fG^1_1 = (wrr'-r)',$$

bearing in mind that here $G^1_1 = G^4_4$ of necessity. It follows that

$$wrr'-r = \tfrac{1}{2}\int r^2 fG^1_1 dw.$$

In view of (9.14) and (9.15) the left hand member has the fixed value -1. Since $G^1_1 = 0$ when $w > -1$ the factor rf in the integrand may be replaced by its value -2 at the origin. The form of G^1_1 must therefore be chosen so that

$$\int rG^1_1 dw = 1. \tag{10.14}$$

Since, from (9.14), $r \sim [2(w+1)]^{1/2}$ as $w \to -1$, this means that

$$G^1_1 = 2[2(1+w)]^{-1/2}\delta(1+w)$$

is an appropriate form of G^1_1. Next, using the relations $2r' = -f, 2r'' = -f'$, equation (9.7) may be recast into

$$\left(r^2 wf'/f + r^2\right)' = r^2 fG^2_2,$$

and $G^2_2 = G^3_3$, of course. Proceeding as before, one must therefore require that $\int rG^2_2 dw = +\tfrac{1}{2}$, or, by comparison with (10.14), $G^2_2 = -\tfrac{1}{2}G^1_1$. The required singular energy tensor is thus finally

$$T^i_j = \kappa^{-1}[2(1+w)]^{-1/2}\delta(1+w)\,\text{diag}\,(2, -1, -1, 2). \tag{10.15}$$

This result seems to me to be of purely formal interest, for the meaningful physical interpretation of singular "sources" of any kind is elusive. Nevertheless one should not fail to note that T^i_j is zero except at the single *instant* $w = -1$.

I conclude this lecture by pointing out that the substitution $r=(1+b/4\bar{r})^2\bar{r}$ converts (10.1) to the so-called isotropic form

$$ds^2 = -(1+\zeta)^4(d\bar{r}^2+\bar{r}^2d\omega^2)+(1-\zeta)^2(1+\zeta)^{-2}dt^2, \quad (10.16)$$

where $\zeta := b/4\bar{r}$. The further substitution $x^1 = \bar{r}\sin\theta\cos\phi$, $x^2 = \bar{r}\sin\theta\sin\phi$, $x^3 = \bar{r}\cos\theta$, $x^4 = t$ converts this into

$$ds^2 = -(1+\zeta)^4\delta_{ab}dx^a dx^b + (1-\zeta)^2(1+\zeta)^{-2}(dx^4)^2.$$

When $\zeta \ll 1$ this is approximated by

$$ds^2 = -(1+b/\bar{r})\delta_{ab}dx^a dx^b + (1-b/\bar{r})(dx^4)^2. \quad (10.17)$$

Note that as $\bar{r}$ goes over the range $0<r<\infty$, r covers the range $b<r<\infty$ twice, and there are no real values of $\bar{r}$ corresponding to the range $0<r<1$. To overcome this hiatus one must drop the assumption that the isotropic metric is independent of x^4.

LECTURE 11

Although spacetimes which deviate greatly from Euclidicity are much the most interesting, I return briefly to the linear approximation where we left off at the beginning of the ninth lecture. First we direct our attention again to astronomical situations and therefore take T^{ij} to represent a static distribution of matter S of sufficiently low density ρ. We go further than this and suppose all motions to proceed with speeds negligible compared with that of light, so that the only surviving component of T^{ij} is $T^{44} = \rho$. This assumption is mandatory if conflict with the assumptions underlying the linear approximation is to be avoided. To illustrate this point clearly, let S be a fluid, so that here $T^{ij} = \text{diag}(p, p, p, \rho)$—recall equation (8.2). Then the linearized equation $T^{ij}{}_{,j} = 0$, or here $T^{ia}{}_{,a} = 0$ reduces to $p_{,a} = 0$. Of course the pressure cannot be everywhere constant and consistency requires that we assume it to be formally negligible in the present context. Writing (8.7) in terms of h_{ij} and bearing in mind that here $\Box$ reduces to $-\nabla^2$, we have

$$\nabla^2 h_{ij} = 2\kappa\left(T_{ij} - \tfrac{1}{2}\eta_{ij}T\right) = \kappa\rho\,\text{diag}(1,1,1,1). \tag{11.1}$$

Let S be spherically symmetric and $\rho = 0$ for $r > R$, where $r^2 := \delta_{ab}x^ax^b$. Then at exterior points, that is, for $r > R$,

$$h_{ij} = -(\kappa m/4\pi r)\,\text{diag}(1,1,1,1),$$

where

$$m := 4\pi \int_0^R \rho r^2 dr. \tag{11.2}$$

If we write $b = \kappa m/4\pi$ here,

$$dr^2 = -(1+b/r)\delta_{ab}dx^adx^b + (1-b/r)(dx^4)^2. \tag{11.3}$$

This must surely be consistent with (10.1) as $r \to \infty$, though, of course, the variable r in (10.1) and the r in (11.3) need not be the same. They need coincide only in the limit $r \to \infty$. Comparison of (11.3) with (10.17) at once reveals that consistency in fact obtains. The parameter b in the sperically symmetric vacuum solution which forms smoothly onto the solution representing S has now received an interpretation: to within a universal constant factor $4\pi/\kappa$ it is the inertial mass of S, defined by (11.2). We must certainly bear in mind, however, that we are only operating within the framework of the linear approximation.

In the case of sufficiently slowly moving matter the retardation in (9.1) can be disregarded as a first approximation and then (11.1) applies, with ρ a slowly varying function of x^4. Under these circumstances one has

$$ds^2 = -(1+2\Omega)\delta_{ab}dx^a dx^b + (1-2\Omega)(dx^4)^2 \qquad (11.4)$$

where Ω satisfies the equation

$$\nabla^2\Omega = -\tfrac{1}{2}\kappa\rho. \qquad (11.5)$$

To find the motion of a test particle the timelike geodesics of (11.4) must be found. Granted that the motion is sufficiently slow $|\dot{x}^a| \ll \dot{x}^4 \approx 1$ and the first three of the geodesic equations reduce to $\ddot{x}^a + \Gamma^a{}_{44} = 0$. Here $\Gamma^a{}_{44} = -\tfrac{1}{2}\eta^{ab}h_{44,\,b} = -\Omega_{,a}$, so that

$$\ddot{x}^a = \Omega_{,a}. \qquad (11.6)$$

Now, astronomers have traditionally determined the motions of celestial objects from equations of the form

$$\nabla^2\Omega = -4\pi k\rho, \qquad d^2x^a/dt^2 = \Omega_{,a},$$

where k is a universal constant called Newton's constant. These equations have exactly the form of (11.5), (11.6) and we can therefore identify κ with $8\pi k$. Then $b = \kappa m/4 = 2km$. As all along, distances are still measured in seconds; and we shall not forgo the convenience of choosing a unit of mass in such a way that the value of k becomes unity. Then $\kappa = 8\pi$ and $b = 2m$.

It seems, then, that the motion of any particular small slowly moving particle is, in the lowest order of approximation, to be found by calculating Ω according to (11.5) from the instantaneous mass distribution of all other like bodies present and then determining the appropriate timelike geodesic of the metric (11.4). This is the heuristic trick of which I spoke last time. The argument is hardly secure, but it is pretty conclusively supported by the fact that, to the order in question, all known approximative schemes for the solution of the field equations lead to the same conclusion.

Having repeatedly referred to the type of spherically symmetric static metric belonging to a material distribution S surrounded by a vacuum, the time has come to examine such metrics and the physics they represent in greater detail. Of course we no longer suppose the metric to be nearly Euclidean overall. First of all, S will be taken to be a fluid so that the interior stresses reduce merely to a hydrostatic pressure and we have once again the simple energy tensor

$$T_i^j = \mathrm{diag}\,(-p, -p, -p, \rho). \tag{11.7}$$

As we noted previously, consistency demands that all thermodynamic fluxes be absent, that is to say, the fluid must be adiabatic. This entails that its equation of state is a relation between p and ρ alone. The g-T pairs now contemplated are—in principle at any rate—intended to reflect physically admissible situations. Various ancillary conditions must therefore be satisfied, one of which is that the metric and the energy tensor be free of singularities. We shall deal with others as we come to them.

For the purpose of investigating static, spherically symmetric nonvacuum solutions we begin with a metric of the generic form (9.5). There we found the presence of an event horizon outside which the use of the manifestly asymptotically Euclidean metric (10.1) proved advantageous. In the case under consideration we shall find that event horizons cannot occur and it is natural to take from the outset a generic form of the metric which includes (10.1) as an immediate special case. Accordingly, rewrite the general spherically symmetric metric (9.5) as

$$ds^2 = -\left(A^{1/2}dR + A^{-1/2}Bdx^4\right)^2 + \left(C + B^2/A\right)\left(dx^4\right)^2 - Dd\omega^2.$$

The linear differential form $A^{1/2}dr + A^{-1/2}B\,dx^4$ can always be written as $e^{\lambda/2}\,dr$ where λ and r are functions of R and x^4. At the same time, write $C + B^2/A =: e^\nu$, $D =: r^2 e^\mu$, $x^4 =: t$ and adopt r as a coordinate in place of R. Then

$$ds^2 = -e^\lambda\,dr^2 - r^2 e^\mu\,d\omega^2 + e^\nu\,dt^2, \tag{11.8}$$

where λ, μ, and ν are functions of r and t. (11.8) will be static if they are in fact independent of t and we now take this condition to be fulfilled. It is advisable to retain both λ and μ explicitly for the time being despite the fact that, because one still has the freedom to make a substitution of the form

$r = f(\bar{r})$, one or other of the functions λ and μ can in effect be freely prescribed, except that one cannot take $\mu = -2\ln r$. The generic Schwarzschild form and isotropic form of the metric correspond to the choices $\mu = 0$ and $\mu = \lambda$, respectively.

The only nonvanishing components of G_{ij} turn out to be the following:

$$G_{11} = \tfrac{1}{4}\mu'^2 + \tfrac{1}{2}\mu'\nu' + r^{-1}(\mu' + \nu') + r^{-2}(1 - e^{\lambda - \mu}),$$

$$G_{22} = \tfrac{1}{2}r^2 e^{\mu - \lambda}\Big[\mu'' + \nu'' - \tfrac{1}{2}\lambda'\mu' - \tfrac{1}{2}\lambda'\nu' + \tfrac{1}{2}\mu'^2 + \tfrac{1}{2}\mu'\nu'$$

$$+ \tfrac{1}{2}\nu'^2 + r^{-1}(2\mu' - \lambda' + \nu')\Big],$$

$$G_{33} = \sin^2\theta\, G_{22},$$

$$G_{44} = -e^{\nu - \lambda}\Big[\mu'' - \tfrac{1}{2}\lambda'\mu' + \tfrac{3}{4}\mu'^2 + r^{-1}(3\mu' - \lambda') + r^{-2}(1 - e^{\lambda - \mu})\Big],$$

$$(11.9)$$

primes denoting derivatives with respect to r. The equations $G_i{}^j{}_{;j} = 0$ reduce to the single equation

$$\left(G_1{}^1\right)' + \left(\mu' + 2/r\right)\left(G_1{}^1 - G_2{}^2\right) + \tfrac{1}{2}\nu'\left(G_1{}^1 - G_4{}^4\right) = 0. \quad (11.10)$$

Now suppose we simply choose the form of the functions λ and μ. The $G_i{}^j$ then become functions of ν'', ν' and r alone, with $G_1{}^1$ in particular not involving ν''. Since $G_i{}^j = 8\pi T_i{}^j$, equation (11.10) then simply amounts to a second order differential equation for ν and the whole problem seems easy enough. Alas, this is very far from being the case. As I already indicated in general terms about halfway through the eighth lecture, the trouble with this scheme is that it almost always leads to functions p and ρ which make little or no physical sense: p or ρ or both may be negative for some values of r; p may fail to vanish for finite values of r; p or ρ may be increasing functions of r; ρ may decrease with p; and so on: and none of these possibilities are acceptable. If S is to be gaseous the position is worse, for it would be almost miraculous to find that p and ρ decrease with increasing r just so as to vanish together at $r = r_0$. In short, the method under discussion is to all intents and purposes useless. What are we to do?

Now, much of astrophysics deals with gaseous stars and it would be instructive to find at least one solution of the equation $G_{ij} = 8\pi T_{ij}$ representing such an object S. If this model solution is to be exact we shall no doubt have to allow some gross oversimplifications. Most obviously, (11.7) already implies that S does not radiate. Still, we may think of S being in a condition such that the dynamical implications of radiation may to a first approximation be neglected altogether. What suggests itself, then, is that one should adopt some suitable equation of state $p = p(\rho)$, just as one does in elementary astrophysics when polytropic distributions are investigated for which $p\rho^{-(1+1/n)} = \text{constant}$, where $n > 0$ is the constant "polytropic index." This proposal, too, is unrealistic in practice; it leads to quite intractable equations. True, there is one particular equation of state, namely $\rho = \text{constant}$, which is an exception to the rule for it leads easily to the so-called Schwarzschild interior solution to which I shall return during the next lecture. Not surprisingly this is, almost without exception, the only static nonvacuum solution considered in the textbooks. For the present we must, however, reject it. Not only is the constancy of ρ repellent in the first place, but it leads to the violation of causality because the speed of sound

$$v = (dp/d\rho)^{1/2} \qquad (11.11)$$

is everywhere infinite. Furthermore, our requirement that ρ vanish on the boundary of S obviously cannot be satisfied.

The most difficult condition to accommodate is the vanishing of ρ on the boundary. This suggests a strategy which ensures from the outset that p and ρ have a common factor which might finally be responsible for a common zero of p and ρ. Since λ, μ are functions of r alone we can always write $\lambda = f(\mu), \nu = g(\mu)$, granted that $\mu \neq 0$. Then, if dots and primes denote derivatives with respect to μ and r respectively, the equations to be satisfied are, from (11.7) and (11.9),

$$p = e^{-f}\left[\left(\tfrac{1}{2}\dot{g} + \tfrac{1}{4}\right)\mu'^2 + r^{-1}(\dot{g}+1)\mu' + r^{-2}(1 - e^{f-\mu})\right], \qquad (11.12)$$

$$p = e^{-f}\left[\tfrac{1}{2}(\dot{g}+1)\mu'' + \tfrac{1}{4}(2\ddot{g} + \dot{g}^2 - \dot{f}\dot{g} + \dot{g} - \dot{f} + 1)\mu'^2 + \tfrac{1}{2}r^{-1}(\dot{g} - \dot{f} + 2)\mu'\right], \qquad (11.13)$$

$$\rho = -e^{-f}\left[\mu'' + \left(\tfrac{3}{4} - \tfrac{1}{2}\dot{f}\right)\mu'^2 + r^{-1}(3 - \dot{f})\mu' + r^{-2}(1 - e^{f-\mu})\right]. \qquad (11.14)$$

Here a constant 8π has been absorbed in ρ and p. In passing, these must identically satisfy (11.10), that is,

$$p' + \tfrac{1}{2}\nu'(p + \rho) = 0, \qquad (11.15)$$

and this is occasionally called the "equation of hydrostatic equilibrium."

The presence of the last terms on the right of (11.12) and (11.14) is inconvenient and we might seek to remove them by choosing isotropic coordinates $f(\mu)=\mu$. Now add p as given by (11.13) to $\frac{1}{2}p$ as given by (11.12) and compare the resulting expression for $\frac{3}{2}p$ with that for $\frac{1}{2}(\dot{g}+1)\rho$. One finds at once that p and ρ have a common factor provided $\ddot{g}+\frac{1}{2}\dot{g}^2+\frac{1}{4}\dot{g}$ $=0$. I won't pursue this case since it leads to an analogue of the polytrope of index 5. Like the latter it extends to infinity and, though interesting in itself, it is not what we want.

Suppose, then, that $f \neq \mu$. From (11.12) and (11.14)

$$-\tfrac{1}{2}(\dot{g}+1)(\rho+p)=\tfrac{1}{2}(\dot{g}+1)e^{-f}\Big[\mu''-\tfrac{1}{2}(\dot{f}+\dot{g}-1)\mu'^2-r^{-1}(\dot{f}+\dot{g}-2)\mu'\Big].$$

$$(11.16)$$

If one now identifies in (11.13) and (11.16) the factors multiplying μ'^2 with each other and likewise the factors multiplying μ', then once again ρ and p will have a common factor. This identification requires that

$$\ddot{g}+\dot{g}^2+\tfrac{1}{2}\dot{g}=0 \qquad \text{and} \qquad \dot{g}(\dot{g}+\dot{f})=0.$$

$\dot{g}=0$ leads to an overall Euclidean metric which is of no interest. Consequently we have the solutions

$$e^{g}=be^{-\mu/2}-a, \qquad e^{f}=ce^{-g}$$

of the equations just written down, where a, b, c are constant of integration. Equating the alternative expressions for p given by (11.12) and (11.13) one now gets an equation for μ. If $\xi:=2ae^{\mu/2}-b$, it is

$$r^2\big(\xi\xi''-\xi'^2\big)-\xi^2=-\beta^2, \qquad (11.17)$$

where $\beta:=(b^2-4ac)^{1/2}$ is required to be real. It has the solutions

$$\xi=\pm\beta W, \qquad W=\begin{cases} \sin(Ar+B)/Ar, \\ \sinh(Ar+B)/Ar, \\ 1+B/r, \end{cases} \qquad (11.18)$$

where A and B are constants of integration. In terms of ξ one has

$$p=-(a/c)\xi(b+\xi)^{-2}(\xi''+2r^{-1}\xi')$$

$$\rho=-(a/c)(2b-3\xi)(b+\xi)^{-2}(\xi''+2r^{-1}\xi'). \qquad (11.19)$$

Incidentally

$$\chi := 3p/\rho = 3\xi/(2b-3\xi), \tag{11.20}$$

and

$$e^\lambda = c(b+\xi)/a(b-\xi), \qquad e^\mu = (b+\xi)^2/4a^2, \qquad e^\nu = a(b-\xi)/(b+\xi). \tag{11.21}$$

The third of the solutions (11.18) is the singular integral of (11.17). It makes $\rho = p = 0$ everywhere, so that it must be equivalent to the Schwarzschild solution (10.1). Take $\xi = \beta(1 + B/r)$ and let $r \to \infty$. If the metric is to reduce to the usual Euclidean form $ds^2 = -dr^2 - r^2\,d\omega^2 + dt^2$, (11.21) shows that we must have

$$a = k, \qquad b = k+1, \qquad c = 1, \qquad \beta = k-1, \tag{11.22}$$

where k is a positive constant. With these, (11.21) becomes

$$e^{-\lambda} = e^\nu = (1-k\phi)/(1+\phi), \qquad e^\mu = (1+\phi)^2. \tag{11.23}$$

with $\phi := (k-1)B/2kr$. If one now makes the substitution $\bar{r} = r + (k-1)B/2k$ and then drops bars, one gets the vacuum metric (10.1) with $\gamma = 1 - 2M/r$. Here M is defined as follows:

$$M := (k^2 - 1)B/4k. \tag{11.24}$$

If values at the boundary and the center of S are distinguished by subscripts b and c respectively, (11.23) is of course relevant only to the region $r \geq r_b$. To find the form of the metric when $r \leq r_b$ we must investigate the other two solutions given by (11.18). For both of these $\xi'' + 2r^{-1}\xi'$ differs from ξ only by a constant factor. The first zero of p as one goes outwards from the center of S defines its boundary, so that, by inspection of (11.19), the boundary corresponds to a zero of ξ. Therefore, incidentally, from (11.21)

$$(e^\lambda)_b = c/a, \qquad (e^\mu)_b = (b/2a)^2, \qquad (e^\nu)_b = a. \tag{11.25}$$

When $B \neq 0$ ξ has a pole at the origin and then $\chi_c = -1$. This violates the restriction that p and ρ are to be nonnegative. Therefore B must be zero. $r^{-1}\sinh Ar$, however, has no real zeros, so that the second of the solutions (11.18) must be rejected. This leaves only the first solution to be considered.

For it $\xi_c = \pm\beta, \xi_b = 0$, so that from (11.21) $(e^\nu)_c/(e^\nu)_b = (b\mp\beta)/(b\pm\beta)$. The stability of S certainly demands that $p' \leq 0$. Then, since p and ρ must not be negative, (11.15) shows that $(e^\nu)_c/(e^\nu)_b < 1$, from which it follows that ξ_c must be positive. Thus finally

$$\xi = \beta(\sin Ar)/Ar, \qquad (11.26)$$

while b must be positive if χ_c is to be positive. Altogether, in view of (11.25), $a>0, b>0, c>0$. We are encouraged by the form of (11.26), since the function on the right plays a prominent part in the theory of the classical polytrope of index 1.

It remains to fit the interior solution, $r<r_b$, to an exterior solution, $r\geq r_b$. I say "a solution" rather than "the solution" since the exterior metric is determined only to within a transformation of the coordinate r. Still, we are naturally inclined to take it in the form (11.23) and require that the g_{ij} and their first derivatives be continuous across the boundary. These are so-called junction conditions which one has to relax to some extent on certain occasions. We remind ourselves that for $r\geq r_b$

$$\phi = [2M/(k+1)]r^{-1}. \qquad (11.27)$$

The continuity of λ, μ, and ν at the boundary at once gives

$$a = (1-k\phi_b)/(1+\phi_b), \qquad b = 2(1-k\phi_b), \qquad c = 1, \qquad (11.28)$$

from (11.25), together with (11.23), at $r=r_b$. Since $\lambda' + \nu' = 0$ for all values of r, continuity of the derivatives of λ, μ, and ν gives only two further conditions. They turn out to be

$$\beta = b\phi_1/(1+\phi_1), \qquad \beta = b(k+1)\phi_1/[2(1+\phi_1)(1-k\phi_1)].$$

From these one has immediately $\phi_b = (1-k)/2k$ and $\beta = 1-k$. The various constants now become simple functions of k alone:

$$a = k, \qquad b = 1+k, \qquad c = 1, \qquad \beta = 1-k. \qquad (11.29)$$

Recall that these relate to the interior of S, whereas (11.22) related to the exterior region. Also

$$M/r_b = (1-k^2)/4k. \qquad (11.30)$$

Here one sees quite clearly that certainly $0<k\leq 1$, but we shall meet stronger limitations on k in a moment.

It is a particularly appealing feature of the present solution that the equation of state of the material of S can be exhibited in simple closed form. If h is any function of r write $\hat{h}:=h/h_c$. Then from (11.19),(11.26), and (11.29),

$$\hat{p}=4NW^2, \qquad \hat{\rho}=4NW[2(1+k)-3(1-k)W]/(5k-1),$$

where

$$N:=\left[(1+k)+(1-k)W\right]^{-2}, \qquad W=(\pi r/r_b)^{-1}\sin(\pi r/r_b).$$

Eliminating W one gets the remarkably simple equation of state

$$(5k-1)\hat{\rho}=4\sqrt{\hat{p}}-5(1-k)\hat{p}. \qquad (11.31)$$

We saw previously that $k>0$. Actually this limitation is much too weak. Since ρ must not be negative, $k>\tfrac{1}{5}$. Next, in view of (11.11) and (11.31), the speed of sound—the isentropic speed, since heat flow was excluded—is given by

$$v^2=(1-k)\sqrt{\hat{p}}/\left[2-5(1-k)\sqrt{\hat{p}}\right]. \qquad (11.32)$$

This takes its largest value at the center,

$$v_c^2=(1-k)/(5k-3).$$

Certainly v_c^2 must be positive which requires that $k>\tfrac{3}{5}$ and this also ensures that ρ decreases outward for $0<r\le r_b$. Actually causality will be violated unless $v_c\le 1$, so that one has the even stronger inequality $k\ge\tfrac{2}{3}$. It is generally accepted that $\chi\le 1$ though it has occasionally been suggested that under some circumstances only the weaker condition $\chi\le 3$ might operate. However, here $\chi\le\tfrac{3}{7}$ so that one gets nothing new.

All conditions which must be met in principle by a physically admissible g-T pair under the conditions envisaged here have now been met. When $k\to 1$ the equation of state of S becomes $\hat{p}=\hat{\rho}^2$ which is that of a classical polytrope of index 1. On the other hand, when k takes its greatest allowed value $\tfrac{2}{3}$ it is $7\hat{\rho}=12\sqrt{\hat{p}}-5\hat{p}$, so that one has something like a star with a core of slowly varying density surrounded by a polytropic region of index 1. Various physical quantities such as the central pressure and the mean density may be directly calculated from the equations we have obtained. For instance

$$p_c=\pi^2 k(1-k)^2/4r_b^2, \qquad (11.33)$$

where the factor 8π contained in p_c must not be forgotten. Here k may be eliminated in favor of M by means of (11.30). For a given value of r_b the greatest value of p_c corresponds to the smallest allowed value $\frac{2}{3}$ of k, for which $p_c = \pi^2/54 r_b^2$. Bear in mind, however, that r_b is merely the value of r corresponding to the boundary of S—any interpretation in terms of the "radius of S" is subject to the uncertainties surrounding the concept of distance. As for the redshift of light emitted from the surface of S and observed at remote points $z = (e^{-\nu/2})_b - 1 = k^{-1/2} - 1$. Since $k \geq \frac{2}{3}$ no redshift larger than $z = (3/2)^{1/2} - 1 \approx 0.225$ could possibly be observed.

The constant M is the "active mass" of S. From (11.23) one has for sufficiently large r, $e^\nu \sim 1 - (1+k)\phi = 1 - 2M/r$. The orbit of a sufficiently distant test particle is, we know, determined by e^ν alone and $2M$ is that constant which was earlier denoted by b. In other words, M is *defined* in terms of the observed motion of a distant test particle and its value can be calculated from it. What physical system might be present in the region of small r is left entirely open and we must not jump to the conclusion that M is the same as the inertial mass m—this equality we only established under the assumption that spacetime was everywhere nearly Euclidean. Well, we might evaluate the inertial mass m for our S. The amount of matter within the coordinate range dr, $d\theta$, $d\phi$ we take to be that in the elementary cube whose sidelengths are calculated from (10.2). They are $e^{\lambda/2} dr$, $re^{\mu/2} d\theta$, $re^{\mu/2} \sin\theta \, d\phi$. Therefore

$$m = 4\pi \int_0^{r_b} r^2 e^{\mu + \lambda/2} \rho \, dr, \tag{11.34}$$

the integration over angles having been already carried out. Putting the various explicit functions into the integrand, one ends up with an intractable integral. This certainly suggests that M and m can in general not be identified. Also, we surely want to know whether there is perhaps some general result which lies behind the curious smallness of the maximum value of the redshift which we found here. Evidently it is time to abandon our particular solution and deal instead with certain general aspects of solutions representing static fluid spheres.

LECTURE 12

To deal with the questions raised at the end of the preceding lecture we naturally begin with a suitable choice of coordinate system, for we recall that one condition can always be imposed jointly on the functions λ, μ, and ν in (11.8). Inspection of (11.9) shows that the greatest formal simplicity is attained by choosing Schwarzschild coordinates, $\mu=0$, for then the equations for S reduce to

$$r^2 p = e^{-\lambda}(r\nu'+1)-1, \tag{12.1}$$

$$r^2 \rho = e^{-\lambda}(r\lambda'-1)+1, \tag{12.2}$$

$$p' = -\tfrac{1}{2}(\rho+p)\nu'. \tag{12.3}$$

Here the equation $G^2{}_2 = -p$ has been replaced by (11.10). Also, a factor 8π has again been absorbed in ρ and p. The following abbreviations are convenient:

$$w := \tfrac{1}{2}r^{-3}\int_0^r r^2 \rho \, dr, \qquad x := r^2, \qquad y := (1-2xw)^{1/2}, \qquad \zeta := e^{\nu/2}. \tag{12.4}$$

Disallowing any singularity at $r=0$, (12.2) shows at once that $e^{-\lambda}=y^2$. (12.1) becomes

$$p = 4y^2 \zeta_{,x}/\zeta - 2w, \tag{12.5}$$

subscripts following a comma denoting derivatives with respect to the variable in question. (12.3) may be rewritten as

$$p_{,x} = -(4xw_{,x}+6w+p)\zeta_{,x}/\zeta. \tag{12.6}$$

103

Eliminating p between (12.5) and (12.6),

$$(1-2xw)\zeta_{,xx}-(xw_{,x}+w)\zeta_{,x}-\tfrac{1}{2}w_{,x}\zeta=0. \tag{12.7}$$

On defining a new variable $\xi:=\int_0^x dx/y$ this becomes

$$\zeta_{,\xi\xi}-g(\xi)\zeta=0, \tag{12.8}$$

where $g(\xi)$ stands for $w_{,x}$, expressed as a function of ξ.

I shall call S a *regular* sphere if it is finite, free of discontinuities, p and ρ are nonnegative, and ρ does not increase outward. The origin of the last condition is this. Since p and ρ and therefore w are nonnegative, so is $v_{,x}$ according to (12.5). Then (12.3) shows that $p_{,x}$ is nonpositive. Positivity of $w_{,x}$ anywhere would therefore imply an imaginary speed of sound there and that is unacceptable. With $w_{,x}\leq0$, (12.8) shows that $\zeta_{,\xi\xi}\leq0$, whence

$$(\zeta_{,\xi})_c\geq\zeta_{,\xi}\geq(\zeta_{,\xi})_b. \tag{12.9}$$

Now $\zeta_{,\xi}=y\zeta'/2r$ and we need to evaluate the right hand member at the boundary. Write R for the boundary value of r, to avoid confusion with the r_b of the preceding lecture. For $r\geq R$ the vacuum solution is (10.1), with $\gamma\equiv\zeta^2=1-2M/r$. Continuity of ζ and its derivative at $r=R$ therefore requires that $(\zeta_{,\xi})_b=M/2R^3$, bearing in mind that $y_b\zeta_b=1$. Also, $M/R^3=w_b$ directly from the definition of y, taken at $r=R$, together with its continuity there. The inequality $\zeta_{,\xi}\geq\tfrac{1}{2}w_b$ may now be integrated from $r=0$ to $r=R$, so that

$$\Delta-\zeta_c\geq w_b\int_0^R r\,dr/y,$$

where $\Delta:=(\sqrt{g_{44}})_b=(1-2M/R)^{1/2}$. Since $w_{,x}\leq0$, $w\geq w_b$ and in turn $y^2\leq 1-2w_bx$. Consequently

$$\Delta-\zeta_c\geq\tfrac{1}{2}(1-y_b)=\tfrac{1}{2}(1-\Delta). \tag{12.10}$$

The regularity of S requires that $\zeta_c>0$. There follows the important general inequality

$$\Delta>\tfrac{1}{3},$$

valid for any regular sphere at all. Even disregarding all other physical limitations, the redshift of light emitted from S is therefore subject to $z<2$.

We have already encountered the inequality $\chi \leq \alpha$, where α is a number whose value may be left open for the moment. Then

$$p \leq \tfrac{1}{3}\alpha\rho = \tfrac{1}{3}\alpha(4xw_{,x} + 6w) \leq 2\alpha w.$$

Since $y_c = 1$, (12.5) then shows that $(\alpha + 1)w_c \geq 2(\zeta_{,x}/\zeta)_c$. On the other hand, from (12.9), $(\zeta_{,x})_c = (\zeta_{,\xi})_c \geq (\zeta_{,\xi})_b = \tfrac{1}{2}w_b$, the last equality having been established previously. From (12.10) $\zeta_c \leq \tfrac{1}{2}(3\Delta - 1)$. Thus $(\alpha + 1)w_c \geq 2w_b/(3\Delta - 1)$, and upon writing $\delta := w_b/w_c$ we finally have

$$\Delta \geq \tfrac{1}{3}\left[1 + 2\delta/(\alpha + 1)\right]. \tag{12.11}$$

For matter under everyday conditions the trace of the energy tensor T is effectively just ρ. Under extreme conditions of very large temperature or pressure T/ρ goes to zero for ordinary matter and T is exactly zero for electromagnetic radiation. It will therefore suffice to take $\alpha = 1$, in which case (12.11) reduces to $\Delta \geq \tfrac{1}{3}(1 + \delta)$. About the value of δ one cannot say very much without detailed calculation except, of course, that $0 < \delta \leq 1$. Very crudely, $\delta = \tfrac{1}{2}$ might perhaps be accepted as a representative value, and then $\Delta \geq \tfrac{1}{2}$, $z \leq 1$. The fact that distant galaxies exhibit redshifts greater than this — indeed greater than $z = 2$ — perhaps has important implications as regards the form of the metric on a cosmical scale.

The inequalities which we have found may be refined in various ways without the imposition of further conditions such as $dp/d\rho \leq 1$. For example, with $\alpha = 1$, one improved inequality is $\Delta > \tfrac{1}{3}(1 + \sqrt{\delta})$. However, this is not the occasion to consider such refinements, nor to obtain inequalities governing other quantities such as the central pressure. What does remain to be done is to consider the question of the inertial mass m and the active mass M of S.

The relation $w_b = M/R^3$, already derived, immediately gives

$$M = \tfrac{1}{2}\int_0^R r^2 \rho \, dr, \tag{12.12}$$

the factor 8π being still contained in ρ. On the other hand, from (11.34), since $\mu = 0$ now,

$$m = \tfrac{1}{2}\int_0^R r^2 y^{-1}\rho \, dr. \tag{12.13}$$

Since $y \leq 1$ we conclude that $M < m$ always. If $\mu(r) := \frac{1}{2}\int_0^r r^2 \rho \, dr$,

$$M - m = \int_0^R \left\{ 1 - \left[1 - 2r^{-1}\mu(r) \right]^{-1/2} \right\} \mu' dr. \qquad (12.14)$$

In the first approximation $\mu(r)/r$ may be neglected when compared with unity, so that then $M - m = 0$. In the next approximation

$$M - m = -\int_0^R r^{-1}\mu(r) \, d\mu(r). \qquad (12.15)$$

In (12.9) and (12.10) the case of equality obtains only when $w' = 0$, that is, $\rho = $ constant. This particular equation of state we excluded previously because of its unphysical nature in as far as it implies an infinite speed of sound. On the other hand it is often considered for formal reasons, for it gives a particularly simple solution of the equations, called the Schwarzschild interior solution. Here, from (12.4), $y = (1 - \frac{1}{3}\rho r^2)^{1/2}$, $\xi = 6\rho^{-1}(1 - y)$ and, since $g(\xi) = 0$, (12.8) gives at once $\zeta = A + By$, where A and B are constants of integration. Taking the continuity of λ, ν, and ν' at $r = R$ into account, it turns out that $A = \frac{3}{2}\Delta$, $B = -\frac{1}{2}$, so that $\zeta = \frac{1}{2}(3\Delta - y)$. Then, from (12.5)

$$p = \rho(y - \Delta)/(3\Delta - y). \qquad (12.16)$$

The mathematical singularity $\zeta_c = 0$ physically represents infinite central pressure. χ attains its greatest value $3(1 - \Delta)/(3\Delta - 1)$ at the center. When $\chi = \alpha$, $\Delta = (1 + \frac{1}{3}\alpha)/(1 + \alpha)$ in harmony with (12.11), since $\delta = 1$ here. Finally M and m may be evaluated explicitly. Trivially $M = \frac{1}{6}\rho R^3$, while

$$m/M = \frac{3}{2}\eta^{-3}\left[\arcsin \eta - \eta(1 - \eta^2)^{1/2} \right], \qquad (12.17)$$

where $\eta^2 := 2M/R$. For $\eta \ll 1$ therefore, $m/M = 1 + \frac{3}{10}\eta^2 + O(\eta^4)$. In the extreme case $\Delta = \frac{1}{3}$ on the other hand, $m/M \approx 1.641$.

One final comment on inequalities which involve R: one must be careful not to read too much into numerical coefficients which appear in them since R has no immediate significance: it is merely the value of a coordinate. It is therefore advisable to eliminate R in favor of the invariant "proper radius" $R^* := \int_0^R (-g_{11})^{1/2} \, dr = \int_0^R dr/y$, for at least this is unaffected by transformations of the radial coordinate. Since $y^2 \leq 1 - 2w_b x$ one has

$$R^* \geq R\eta^{-1}\arcsin \eta = 2M\eta^{-3}\arcsin \eta. \qquad (12.18)$$

In particular, given the general inequality $\Delta > \frac{1}{3}$, that is $M < \frac{4}{9}R$, the least value of the factor multiplying M in (12.18) is approximately 2.938. One therefore has the invariant limitation $M \lesssim 0.3404 R^*$.

In our considerations of spherically symmetric solutions we have hitherto not contemplated the possible presence of an electrostatic field. Partly in order to rectify an error which has often been made in the past, I therefore go on to investigate very briefly spherically symmetric static solutions of the Einstein-Maxwell equations. To begin with, granted a metric of the type (11.8), it is sufficient to accommodate the condition of static spherical symmetry on the electromagnetic potential A_i by taking $A_a = 0$, $A_4 = \phi(r)$. It is not necessary; but by observing the effects on A_i of infinitesimal rotations —recall the ninth lecture—it turns out that quite generally A_a must have a form just such that it can always be "removed" by a suitable electromagnetic gauge transformation. Further, if $\sigma(r)$ is the invariant electric charge density, $j^i = \sigma u^i$ is the current. For matter at rest $u^i = (0,0,0,u^4)$ and if Schwarzschild coordinates are used again, $u^4 = e^{\nu/2}$ since $u_i u^i = 1$. Of the components of f_{ij} only $f_{14} \equiv -f_{41} = \phi'$ does not necessarily vanish. From (7.17) the electromagnetic energy tensor is then

$$T_i^j = u\,\mathrm{diag}\,(1, -1, -1, 1)$$

where $u := \frac{1}{2} e^{-\lambda-\nu}\phi'^2$. Absorbing a factor 8π in u, we can evidently take over equations (12.1) and (12.2) provided p be replaced by $p-u$, and ρ by $\rho^* := \rho + u$, granted that the stresses reduce merely to a hydrostatic pressure, as before. In place of (12.3) we must go back to (11.10). Altogether

$$r^2(p-u) = e^{-\lambda}(r\nu'+1) - 1 \tag{12.19}$$

$$r^2\rho^* = e^{-\lambda}(r\lambda'-1) + 1 \tag{12.20}$$

$$(p-u)' = -\tfrac{1}{2}(\rho+p)\nu' + 4r^{-1}u \tag{12.21}$$

$$(r^2 e^{-(\lambda+\nu)/2}\phi')' = -\sigma r^2 e^{\lambda/2}, \tag{12.22}$$

the last equation being the sole survivor of the set $f^{ij}{}_{;j} = j^i$.

In pursuit of the classical strategy we first take $\sigma = \rho = p = 0$, but allow the possible appearance of a singularity in the solution of the equations. By addition of (12.19) and (12.20) one infers that $\lambda' + \nu' = 0$, that is, $\lambda + \nu = 0$. An additive constant of integration has been rejected, since it can always be removed by a change of scale of the coordinate t. (12.22) becomes $(r^2\phi')' = 0$, so that

$$\phi' = Q_1/4\pi r^2, \qquad (12.23)$$

where Q_1 is a constant of integration. As a matter of convenience it is useful to write $Q := (4\pi)^{-1/2}Q_1$. Since $\lambda + \nu = 0$, u stands for $4\pi\phi'^2$ so that (12.20) becomes $[r(1 - e^{-\lambda})]' = Q^2/r^2$, whence

$$e^{-\lambda} = e^{\nu} = :\gamma = 1 - 2M/r + Q^2/r^2, \qquad (12.24)$$

where M is a constant of integration. With this form of γ the metric is given by (10.1). This, then, is the so-called Reissner-Nordstrøm solution. It is usually spoken of as representing a region containing a "particle" carrying an electric charge Q_1 — in rationalized units as usual — though this is hardly more than a conventional terminology. Note that provided $M > |Q|$ there again exists an event horizon.

Now, on examining the motion of a distant uncharged test particle P one will obviously find the usual elliptical orbits — granted that by choice $e^2 < 1$ again in (10.6) — with M interpreted as an active mass; for as $r \to \infty$ the term Q^2/r^2 in γ will have no effect. It is therefore sometimes asserted that the presence of the charge does not affect the motion of P, presumably on the grounds that M and Q here turned up as mutually independent constants of integration. However, in the absence of a detailed physical picture of a neighborhood of $r = 0$, who is to say that M and Q can in fact be regarded as being mutually independent?

To throw some light on this question we should consider some more or less physically realistic phenomenological system. I take this to be a singularity-free finite distribution of charged fluid. This is governed by equations (12.19)–(12.22). To make them into a closed system one must in principle now add two further relations, such as those expressing the dependence of σ and of p on ρ. The wealth of possible choices alone makes it largely pointless to find particular solutions. Fortunately some interesting general conclusions can be drawn merely by investigating consequences of the conditions which must be satisfied at the boundary of S, that is, at $r = R$.

For $r \geq R$ the metric is given by (10.1) and (12.24) together. The previous definitions (12.4) may be retained, provided ρ is replaced by ρ^* in them. Further, write $q(r) := 4\pi\int_0^r r^2\sigma e^{\lambda/2}dr$, so that the total charge on S is $Q_1 = q(R)$. Now (12.20) may be written as $[r(1 - e^{-\lambda})]' = r^2\rho^*$. Integrating this from $r = 0$ to some value of $r > R$,

$$\left[r(1 - e^{-\lambda})\right]_0^r = \int_0^R r^2\rho\, dr + \int_0^R r^2 u\, dr + \int_R^r r^2 u\, dr. \qquad (12.25)$$

At the origin $e^\lambda = 1$, since ρ^* is finite there. Using (12.24), the left hand member of (12.25) is therefore equal to $2M - Q^2/r$, granted that γ is continuous on the boundary. On the right, define $M_\rho := \frac{1}{2}\int_0^R r^2 \rho \, dr$ and $M_\sigma := \frac{1}{2}\int_0^R r^2 u \, dr$. The last term can be evaluated as before and becomes $Q^2/R - Q^2/r$. Thus (12.25) gives the relation

$$M = M_\rho + M_\sigma + Q^2/2R. \tag{12.26}$$

This may be further simplified. By inspection of (12.22), $4\pi r^2 \phi' = -q$, so that $u = q^2/4\pi r^4$. Hence, integrating by parts,

$$M_\sigma = \int_0^R (q^2/8\pi r^2) \, dr = -Q^2/2R + \int_0^R (qq'/4\pi r) \, dr.$$

It thus follows that

$$M = M_\rho + \int_0^R (4\pi r)^{-1} q \, dq. \tag{12.27}$$

Clearly the presence of a charge distribution within S implies a well-defined positive contribution to the active mass of S; which is what we set out to show.

The time has come to abandon spherical symmetry and to go on very briefly to axially symmetric static metrics. By this I mean here that any such metric, apart from being static in the usual sense, is also invariant under the transformations $x^3 \to x^3 + \epsilon$ and $x^3 \to -x^3$, where x^3 is a spacelike coordinate and ϵ an arbitrary constant. Therefore, generically,

$$ds^2 = g_{11}(dx^1)^2 + 2g_{12}\, dx^1\, dx^2 + g_{22}(dx^2)^2 + g_{33}(dx^3)^2 + g_{44}(dx^4)^2, \tag{12.28}$$

with g_{ij} independent of x^3 and x^4. By means of a transformation involving x^1 and x^2 alone this can be further simplified to

$$ds^2 = g_{11}\left[(dx^1)^2 + (dx^2)^2\right] + g_{33}(dx^3)^2 + g_{44}(dx^4)^2.$$

Of course g_{11} is not the original function g_{11}. I shall not go through the tedious exercise of writing down the components of the Ricci tensor for this metric. What in fact emerges is that the vanishing of $R^3{}_3 + R^4{}_4$ requires that $r_{,11} + r_{,22} = 0$ where $r^2 := -g_{33}g_{44}$, that is to say, r is a harmonic function of x^1 and x^2. There therefore exists a function $z(x^1, x^2)$ such that $r + iz$ is an analytic function of $x^1 + ix^2$. Now introduce r and z as coordinates in place of x^1 and x^2. Then $g_{11}[(dx^1)^2 + (dx^2)^2]$ becomes $dr^2 + dz^2$ times a function, $-\alpha^2$, say, of r and z. Since $g_{33} = -r^2/g_{44}$, the final form of the metric contains only two unknown functions, namely α and g_{44}. Partly from hindsight, it is advisable to take its generic form to be

$$ds^2 = -e^{2(\lambda-\nu)}(dr^2 + dz^2) - r^2 e^{-2\nu}d\phi^2 + e^{2\nu}dt^2, \qquad (12.29)$$

with ν and λ functions of r and z only. This is usually referred to as the Weyl metric.

(12.29) is the canonical form of the axially symmetric static metric; but bear clearly in mind—it is often forgotten—that the reduction of (12.28) to (12.29) is possible only if $R^3_3 + R^4_4 = 0$. This will certainly be the case for vacuum solutions but not, in general, when $T^{ij} \neq 0$. The vacuum equations will be satisfied provided

$$\nu_{,rr} + r^{-1}\nu_{,r} + \nu_{,zz} = 0, \qquad (12.30)$$

$$\lambda_{,r} = r\left(\nu_{,r}^2 - \nu_{,z}^2\right), \quad \lambda_{,z} = 2r\nu_{,r}\nu_{,z}, \qquad (12.31)$$

$$\lambda_{,rr} + \lambda_{,zz} + \nu_{,r}^2 + \nu_{,z}^2 = 0. \qquad (12.32)$$

Evidently the two equations (12.31) must satisfy an integrability condition, but the latter is just (12.30) again. The remarkable feature of the latter is that it is the ordinary three-dimensional Laplace equation for a function which is independent of ϕ. Further, (12.32) is already implied by the other equations.

A remarkably simple state of affairs has revealed itself—it seems too good to be true. An elementary solution which has a suitable behavior at infinity is

$$\nu = -m/\rho, \qquad \lambda = -m^2 r^2/2\rho^4, \qquad (12.33)$$

with $\rho = (r^2 + z^2)^{1/2}$ and m an arbitrary constant. It has a singularity at the origin, which we may possibly think of as representing a "particle," but this particle will have some kind of structure, in as far as (12.33) does not imply spherical symmetry. To have a metric which can be transformed into the Schwarzschild metric one has to start with a much more complicated solution of (12.30).

Since (12.30) is linear, one can add any number of particular solutions to get a new solution. Take a particular example. For $i = 1, 2$, write $\rho_i := [r^2 +$

$(z-z_i)^2]^{1/2} \geq 0$, where the z_i are given constants. Then, with positive constants m_1, m_2,

$$v = -m_1/\rho_1 - m_2/\rho_2 \qquad (12.34)$$

is a superposition of elementary solutions of the kind (12.33). From (12.31) in turn

$$\lambda = -\tfrac{1}{2}r^2\left(m_1^2/\rho_1^4 + m_2^2/\rho_2^4\right) - 2m_1 m_2 (z_1 - z_2)^{-2}$$

$$\times \left\{1 - \left[r^2 + (z-z_1)(z-z_2)\right]/\rho_1\rho_2\right\}. \qquad (12.35)$$

(12.34) and (12.35) together represent a so-called Curzon solution. At first sight it looks acceptable enough. There are singularities at $z=z_1$ and $z=z_2$ and can one not therefore interpret the solution as representing two isolated particles P_1 and P_2? No, one cannot: two isolated particles, initially at rest, must move toward each other, as we already know. This apparent conflict at one time gave rise to much polemical debate.

To resolve it, consider a very small circle surrounding the z-axis in the plane $z=$ constant, $t=$ constant. Its radius ϵ corresponds to $r=e^{v-\lambda}\epsilon$, whereas its circumference is $2\pi re^{-v} = 2\pi e^{-\lambda}\epsilon$. In a sufficiently small neighborhood of any regular point every metric must be Euclidean, so that the circumference of the circle must be $2\pi\epsilon$. We conclude that λ must vanish at $r=0$. According to (12.35), however, the value of λ is zero only if z does not lie between z_1 and z_2, being $-4m_1m_2(z_1-z_2)^{-2}$ otherwise. The z-axis is therefore a line singularity between P_1 and P_2. A picturesque physical interpretation of this is that of a strut keeping P_1 and P_2 apart.

That all the explicit vacuum solutions we have encountered so far contain singularities is no accident. Consider any static asymptotically Euclidean metric

$$ds^2 = g_{ab}dx^a dx^b + e^{2q}(dx^4)^2, \qquad (12.36)$$

say. Distinguish by a bar any quantity which belongs to the three-dimensional metric g_{ab}, subscripts following a colon denoting covariant derivatives with respect to $\bar{\Gamma}^c_{ab}$. Then

$$R_{ab} = \bar{R}_{ab} + q_{:ab} + q_{:a}q_{:b}, \qquad R_{44} = e^{2q}(q_{:c}{}^c + q_{:c}q^{:c}). \qquad (12.37)$$

If $R_{ij} = 0$ the second of these equations shows that $(e^q)_{:c}{}^c$ must vanish. This is a linear elliptic differential equation for e^q and one knows of any equation of this type that if a solution asymptotically approaches a constant value and is regular everywhere then it is a constant. The vanishing of R_{ab} now requires that of $\bar{R}_{ab}$. However, the three-dimensional Riemann tensor $\bar{R}_{abcd}$ has only six linearly independent components and the same is true also of $\bar{R}_{ab}$. $\bar{R}_{abcd}$ is therefore related linearly and homogeneously to the $\bar{R}_{ef}$ and so vanishes here. This implies that g_{ab} can be reduced to η_{ab} by a transformation of the coordinates x^c and the metric g_{ij} is now manifestly Euclidean. The upshot of all this is that every regular, static, asymptotically Euclidean solution of $R_{ij} = 0$ is Euclidean. If it is to be otherwise the presence of singularities must be admitted.

The equations (12.37) prompt a small digression. Write $g_{ab} =: e^{-2q}\hat{g}_{ab}$, $\hat{\Gamma}^c_{ab}$ and $\hat{R}_{ab}$ having obvious meanings. If subscripts following a bar denote covariant derivatives with respect to $\hat{\Gamma}$

$$R_{ab} = \hat{R}_{ab} + 2q_{|a}q_{|b} - \hat{g}_{ab}q_{|c}{}^c, \qquad R_{a4} = 0, \qquad R_{44} = e^{4q}q_{|c}{}^c.$$

The equations $R_{ij} = 0$ therefore become

$$\hat{R}_{ab} = -2q_{|a}q_{|b}, \qquad q_{|c}{}^c = 0.$$

Inspection shows these to be unaffected by the substitution $q \to -q$. It follows that if the static metric $ds^2 = g_{ab}\,dx^a\,dx^b + g_{44}\,dt^2$ satisfies the vacuum equations $R_{ij} = 0$ then so does $ds^2 = (g_{44})^2 g_{ab}dx^a dx^b + (g_{44})^{-1}(dx^4)^2$. The timelike character of x^4 is irrelevant to this result which was the first of a long line of methods for generating solutions of one set of equations from known solutions of the same or some other appropriate set.

I have devoted much time to the spherically symmetric static solutions, not only because they provided us with an observational validation of the theory, but also because it was possible to find nonvacuum solutions which bore at least some resemblance to actual physical situations. Some unfamiliar conclusions emerged, such as the limitation on the value of M/R^*. Likewise, unfamiliar features characterize the orbits of test objects when in (10.1) the restriction $r \gg b$ is abandoned, for instance with regard to their stability. The point is that to make exact theoretical predictions one needs to have exact representative solutions of the field equations, representative in the sense that they are physically not totally unrealistic. It seems we should seek such solutions representing systems which lack one or other or both of the properties of being spherically symmetric and static.

Alas, we can do little more than glance at this general problem. Nonstatic spherically symmetric systems—"stars"—are of course of astrophysical interest, but even here success has been only marginal. Moreover, granted that the star is not emitting significant amounts of matter and radiation—on any reasonable time scale this is the case with most stars, as their long lifetimes demonstrate—the vacuum metric in the regions surrounding them must be just the usual static exterior metric. We recall that a test particle outside the star moves exactly as if the star were static. That is not very interesting and we are led to contemplate nonspherically symmetric stars. Nonradial pulsations will imply a nonstatic metric and the problem will almost certainly be intractable. This leaves us only with stationary axially symmetric metrics, representing rotating stars, as possible candidates for likely success. So far this hope has proved illusory as far as interior regions are concerned. We must further whittle down our expectations and restrict our search for vacuum metrics which might be suitable candidates for the exterior region of a steadily rotating star. Such a solution—the so-called Kerr metric—is known.

Before I go on to this—because of its importance I must do so—let me just say again that for many years a tremendous effort has gone into finding explicit vacuum metrics. To this end many special formal schemes have been devised with often quite remarkable ingenuity, though in the end many of the metrics which have so become available seem to lack physical interest. The Kerr metric is a fortunate exception in this respect. Its original derivation rested on the use of a quite complex formal apparatus and even a simplified method devised subsequently still involves relatively lengthy calculations. Therefore, when we come to it next time I shall—without running counter to the general character of these lectures—omit extensive routine steps, such as the evaluation of Christoffel symbols and of components of the Ricci tensor or those involved in making mere changes of variables.

LECTURE 13

Generically the metric of the Kerr solution which we are seeking is axially symmetric, but now merely stationary. This means it is still invariant under $x^3 \rightarrow x^3 + \epsilon$, but no longer under $x^3 \rightarrow -x^3$. The metric (12.28) must therefore be supplemented by an additional term $2g_{34} \, dx^3 \, dx^4$. As before, the sum of the first three terms can in effect always be written $g_{11}[(dx^1)^2 + (dx^2)^2]$. Now let $r^2 := -g_{33}g_{44} + g_{34}^2$. Upon writing down the field equations $R_{ij} = 0$, inspection reveals that r must be a harmonic function of x_1 and x_2. Let z be its conjugate harmonic function. Then, exactly as before, we take r and z as coordinates in place of x^1 and x^2. As a result the metric becomes

$$ds^2 = -\alpha^2(dr^2 + dz^2) + (g_{44})^{-1}\left\{ -\left[r^2 - (g_{34})^2\right](dx^3)^2 + \right.$$

$$\left. + 2g_{34}g_{44} dx^3 \, dx^4 + (g_{44})^2(dx^4)^2\right\}.$$

Now write, with f, γ, and ω functions of r and z alone,

$$x^3 =: \phi, \qquad x^4 =: t, \qquad \alpha^2 =: f^{-1}e^{2\gamma}, \qquad g_{44} =: f, \qquad g_{34}/g_{44} =: -\omega.$$

Then the metric takes its canonical form

$$ds^2 = -f^{-1}\left[e^{2\gamma}(dr^2 + dz^2) + r^2 \, d\phi^2\right] + f(dt - \omega \, d\phi)^2. \qquad (13.1)$$

When $\omega = 0$ this coincides with (12.29). Here also the reduction to canonical form was of course possible only because the field equations are satisfied, at least in part.

One has four field equations, E_1 to E_4, say, which generalize (12.30)–(12.32) on account of the presence of the derivatives of ω, but E_4 is already contained in the other three. There is also a fifth equation—E_5, say—which is nugatory when $\omega = 0$. In terms of an auxiliary three-dimensional Euclidean space endowed with cylindrical polar coordinates r, z, ϕ one has the usual scalar and vector operators. For example, if $\mathbf{A}$ is a vector, $\nabla \cdot \mathbf{A} = r^{-1}(rA_r)_{,r} + A_{z,z}$. Then E_1 and E_5 may be written in the compact form

$$f \nabla^2 f = \nabla f \cdot \nabla f + r^{-2} f^4 \nabla \omega \cdot \nabla \omega, \tag{13.2}$$

$$\nabla \cdot (r^{-2} f^2 \nabla \omega) = 0. \tag{13.3}$$

On account of (13.3) there exists a vector $\mathbf{A}$ such that

$$r^{-2} f^2 \nabla \omega = \nabla \times \mathbf{A}. \tag{13.4}$$

Since the ϕ-component of the gradient on the left vanishes, it follows that $A_{z,r} - A_{r,z} = 0$, so that there exists a function $F(r, z, \phi)$ such that $A_r = F_{,r}$, $A_z = F_{,z}$. In consequence, if $\Phi := F_{,\phi} - rA_\phi$, (13.4) becomes, in component form,

$$(\omega_{,r}, \omega_{,z}, 0) = rf^{-2}(\Phi_{,z}, -\Phi_{,r}, 0), \tag{13.5}$$

from which it follows at once that

$$\nabla \cdot (f^{-2} \nabla \Phi) = 0. \tag{13.6}$$

Now, remarkably, the equations (13.2) and (13.3) can be combined into the single equation

$$(\operatorname{Re} \mathcal{E}) \nabla^2 \mathcal{E} = \nabla \mathcal{E} \cdot \nabla \mathcal{E} \tag{13.7}$$

for the complex function

$$\mathcal{E} := f + i\Phi. \tag{13.8}$$

$\mathcal{E}$ is known as the Ernst potential. To verify this, substitute (13.8) into (13.7), separate real and imaginary parts, and use (13.2), (13.5), and (13.6).

It is of advantage to introduce in place of $\mathcal{E}$ the complex function ξ:

$$\mathcal{E} =: (\xi - 1)/(\xi + 1) \tag{13.9}$$

in terms of which (13.7) becomes

$$(\xi \bar{\xi} - 1) \nabla^2 \xi = 2\bar{\xi} \nabla \xi \cdot \nabla \xi, \tag{13.10}$$

bars denoting complex conjugation. Once one has any solution of this equation the required functions f, ω, and γ are obtained from (13.8), (13.5), and E_2, E_3 together, respectively. In fact

$$f = \text{Re}\left[(\xi - 1)/(\xi + 1)\right],$$

$$\omega_{,r} = -2r(\xi\bar{\xi} - 1)^{-2}\text{Im}\left[(\bar{\xi} + 1)^2\xi_{,z}\right], \quad \omega_{,z} = 2r(\xi\bar{\xi} - 1)^{-2}\text{Im}\left[(\bar{\xi} + 1)^2\xi_{,r}\right],$$

$$\gamma_{,r} = r(\xi\bar{\xi} - 1)^{-2}(\xi_{,r}\bar{\xi}_{,r} - \xi_{,z}\bar{\xi}_{,z}), \quad \gamma_{,z} = 2r(\xi\bar{\xi} - 1)^{-2}\text{Re}(\xi_{,r}\bar{\xi}_{,z}). \quad (13.11)$$

In practice the solution of (13.10) is often facilitated by introducing prolate spheroidal coordinates $r = k(x^2 - 1)^{1/2}(1 - y^2)^{1/2}$, $z = kxy$, where k is an arbitrary constant. All our previous solutions and generalizations of these can thus be recovered, as can many other known solutions. New, however, is the class of so-called Tomimatsu-Sato solutions, for instance. Generically these have $\xi = P(x, y; p, q, \delta)/Q(x, y; p, q, \delta)$, where P and Q are complex polynomials of degree δ^2 and $\delta^2 - 1$, respectively, $\delta = 1, 2, 3, \ldots$ and p, q are constant real parameters with $p^2 + q^2 = 1$. When $\delta = 2$

$$P = p^2x^4 + q^2y^4 - 2ipqxy(x^2 - y^2), \quad \tfrac{1}{2}Q = px(x^2 - 1) - iqy(1 - y^2),$$

so that one already has quite a complicated state of affairs, more so for $\delta > 2$. The case $\delta = 1$ alone is relatively simple and it gives just the Kerr solution, for which

$$\xi = px - iqy.$$

Then it turns out eventually that, with $w := p^2x^2 + q^2y^2 - 1$

$$1/f = \left[(px + 1)^2 + q^2y^2\right]/w, \quad \omega = -2kp^{-1}q(1 - y^2)(px + 1)/w,$$

$$e^{-2\gamma} = p^2(x^2 - y^2)/w.$$

Also, $dr^2 + dz^2 = k^2(x^2 - y^2)[dx^2/(x^2 - 1) + dy^2/(1 - y^2)]$. Finally, write $M := k/p$, $a := kq/p$ and make the transformation $px = (\rho/M) - 1$, $qy = (a/M)\cos\theta$ to new variables ρ, θ. As a result the Kerr metric emerges in the so-called Boyer-Lindquist form

$$ds^2 = -\frac{\rho^2 + a^2\cos^2\theta}{\rho^2 - 2M\rho + a^2}d\rho^2 - (\rho^2 + a^2\cos^2\theta)d\theta^2$$

$$- \left(\rho^2 + a^2 + + \frac{2Ma^2\rho\sin^2\theta}{\rho^2 + a^2\cos^2\theta}\right)\sin^2\theta\, d\phi^2 + \frac{4Ma\rho\sin^2\theta}{\rho^2 + a^2\cos^2\theta}d\phi\, dt$$

$$+ \left(1 - \frac{2M\rho}{\rho^2 + a^2\cos^2\theta}\right)dt^2. \quad (13.12)$$

When $a=0$ the metric is static: it is exactly the Schwarzschild metric (10.1). The charged counterpart to the latter is the Reissner-Nordstrøm solution (12.24). I mention in passing that a solution of the Einstein-Maxwell equations is known which is the charged generalization of (13.12)—the so-called Kerr-Newman metric—which reduces to (13.12) when $Q=0$ and to the Reissner-Nordstrøm metric when $a=0$. To gain some physical insight into (13.12), consider almost Euclidean conditions, that is to say, suppose M/ρ and a/ρ to be so small that their squares can be neglected when compared with unity. Then (13.12) reduces to

$$ds^2 = -(1+2M/\rho)\,d\rho^2 - \rho^2(d\theta^2 + \sin^2\theta\,d\phi^2)$$

$$+4Ma\rho^{-1}\sin^2\theta\,d\phi\,dt + (1-2M/\rho)\,dt^2.$$

Upon making the substitution $\rho=\bar r+M$ and then dropping the bars, this becomes, to the same approximation,

$$ds^2 = -(1+2M/r)\left[dr^2 + r^2(d\theta^2 + \sin^2\theta\,d\phi^2)\right]$$

$$+4Mar^{-1}\sin^2\theta\,d\phi\,dt + (1-2M/r)\,dt^2. \tag{13.13}$$

This we can now compare with the general solution (9.1) of the linearized equations if we take (13.13) to represent the distant vacuum region surrounding a body S of some kind. The generic form of the exact solution (13.12) implies generic properties of this body: it must be axially symmetric and be in steady rotation. Furthermore, it must be symmetric about the plane $\theta=\pi/2$. Consistency with the linear approximation demands that S must be sufficiently diffuse, that all stresses be negligible, and that all speeds be very much less than unity. Under these circumstances the "static approximation" $S_{ij}:=T_{ij} - \frac{1}{2}\eta_{ij}T=\frac{1}{2}\rho\,\mathrm{diag}(1,1,1,1)$ must be supplemented by the inclusion of the off-diagonal elements $S_{a4}=S_{4a}=T_{a4}=\rho u_a$. The only components of h_{ij} which differ from those obtained earlier are therefore the $h_a:=h_{a4}=h_{4a}$. They obey the equation

$$\nabla^2 h^a = 16\pi T^{4a}.$$

Recalling (9.1), this has the solution

$$h^a = 4\int R^{-1}T^{4a}d^3\bar x.$$

When $r\gg\bar r$ the approximation $R^{-1}=r^{-1}+\delta_{bc}x^b\bar x^c r^{-3}$ suffices. The term of h^a corresponding to r^{-1} gives no contribution since the total "linear momentum" $\int T^{4a}d^3\bar{x}$ vanishes. The dominant term of h^a is thus given by

$$h^a = 4\delta_{bc}x^b r^{-3}\int \bar{x}^c T^{4a}d^3\bar{x}.$$

Now, dropping bars for the moment, $(x^a x^c T^{4b})_{,b} = x^c T^{4a} + x^a T^{4c}$ since $T^{4j}_{,j} = T^{4b}_{,b} = 0$. Integrating the divergence over all values of the coordinates it follows that $\int x^c T^{4a}d^3x = -\int x^a T^{4c}d^3x$. Thus

$$h^a = 2\delta_{bc}x^b r^{-3}\int (\bar{x}^c T^{4a} - \bar{x}^a T^{4c})d^3\bar{x}. \qquad (13.14)$$

The total "angular momentum" J of S is, by definition,

$$J_a := \epsilon_{abc}\int \bar{x}^b T^{4c}d^3\bar{x},$$

so that the integral on the right of (13.14) is $-\epsilon^{acd}J_d$. Thus now

$$h_a = -2r^{-3}\epsilon_{abc}x^b J^c. \qquad (13.15)$$

Here, from first principles, only J^3 does not vanish. Writing simply J for this

$$h_1 = -2Jr^{-3}x^2, \qquad h_2 = 2Jr^{-3}x^1, \qquad h_3 = 0. \qquad (13.16)$$

Finally, $2h_a dx^a = 4Jr^{-3}(x^1 dx^2 - x^2 dx^1)$, so that going to polar variables,

$$2h_a dx^a = 4Jr^{-1}\sin^2\theta\, d\phi. \qquad (13.17)$$

Comparing this with the penultimate term of (13.13), one arrives at the identification

$$a = J/M. \qquad (13.18)$$

One must not lose sight of the fact that (13.18) has been the result of an approximate calculation. One has no assurance that if one had an exact interior solution matching (13.12) the intrinsic angular momentum of S—it is not quite clear how one might define it—would turn out to have the value Ma which occurs in (13.12). We remind ourselves that the value of the constant M in (10.1) was not that of what we called the inertial mass. I shall call J the active angular momentum of S. To find it one has to investigate the effects of the presence of the term $2g_{34}\,d\phi\,dt$ in (13.12) on particle tracks and light rays.

To this end we should now write down the explicit geodesic equations. That is easy enough: find the equation $\delta L/\delta \dot{x}^i = 0$ directly from the Lagrangian $L = g_{ij}\dot{x}^i\dot{x}^j$. A general orbit will now not be plane and we cannot willy-nilly put $\theta = \pi/2$. We are still confronted by a formidable problem and we therefore content ourselves with a glance at equatorial orbits, $\theta = \pi/2$, for these will be plane by symmetry. We have the integrals of motion

$$g_{33}\dot{\phi} + g_{34}\dot{t} = -h, \qquad g_{34}\dot{\phi} + g_{44}\dot{t} = c, \qquad L = 1,$$

which correspond to (10.3) and (10.4). Here

$$g_{33} = -(\rho^2 + a^2 + 2Ma^2/\rho), \qquad g_{34} = 2Ma/\rho, \qquad g_{44} = 1 - 2M/\rho.$$

$$(13.19)$$

From the first two it follows that

$$\rho^2\dot{\phi} = \left[h(1 - 2M/\rho) + 2cJ/\rho \right] / (1 - 2M/\rho + a^2/\rho^2).$$

Now suppose a particle P is coming in radially from infinity. The constant of integration h must vanish and, since $c > 0$, $\rho^2\dot{\phi} > 0$. Instead of continuing along the initial radius, P will be deflected as if the central rotating object were dragging it with it—one speaks of frame-dragging.

We can look at this phenomenon also by contemplating circular orbits, $\rho = $ constant, $\theta = \pi/2$. Then the condition for the geodesic to be timelike is

$$g_{33}\Omega^2 + 2g_{34}\Omega + g_{44} > 0, \qquad (13.20)$$

where $\Omega := d\phi/dt$. Ω is the angular velocity of P in the following sense. Let a sufficiently distant observer determine on *his* clock the time T taken for one complete revolution of P about S. Since the spacetime is stationary the corresponding change Δt of t is the same for the observer as for P—recall our discussion of the redshift. For him, however, $\Delta s = \Delta t$ so that, $T = 2\pi/\Omega$. (13.20) now gives, with $D := (\rho^2 - 2M\rho + a^2)^{1/2} > 0$,

$$2J/\rho - D < (\rho^2 + a^2 + 2Ma^2/\rho)\Omega < 2J/\rho + D, \qquad (13.21)$$

J having been taken as positive here. Therefore, when $\rho \to \infty$, $-1 < \rho\Omega < 1$, as should be. However, when $\rho \to 2M$ the least value of Ω is zero: P can no longer rotate about S in a sense opposite to that of the rotation of S itself.

As in the spherically symmetric case one can of course consider the vacuum metric here without thinking in the first place about any realistic distribution of matter to which it might be matched. Inspection of (13.12) then reveals that there are apparently four distinct singular surfaces analogous to the surface $r = b$ of the Schwarzschild metric (10.1), namely $B_{\pm}: \rho = M \pm (M^2 - a^2)^{1/2}$ and $B_{\pm}^*: \rho = M \pm (M^2 - a^2 \cos^2\theta)^{1/2}$, subject to the assumption that $M > |a| \neq 0$. These, like the singularity at $\theta = 0$, are however, spurious. The Kerr spacetime has only one genuine singularity. This occurs at $\rho = 0$: as $\rho \to 0$ the invariant $R_{ijkl}R^{ijkl}$ diverges. As before one can draw sensible conclusions from (13.12) about the region lying within B_+^*, provided one takes into account the changes of character of coordinates from timelike to spacelike and conversely. B_+^* is formally a surface of infinite redshift at which, moreover, frame-dragging becomes complete; but, ignoring the points $\theta = 0, \pi$, it is not an event horizon. That role is reserved for the surface B_+. I leave B_-^* and B_- aside since they lie entirely within B_+.

Evidently the Kerr metric represents a black hole when $M^2 \geq Q^2 + a^2$. The Kerr-Newman metric formally differs from (13.12) only in that $2M - Q^2/\rho$ replaces $2M$: and it also is a black hole solution. The latter is remarkable in as far as the following seems to be true: if an actual star, no matter how complex the motion of its parts, collapses to become occluded, then in its final state it will be described by the Kerr-Newman metric. Therefore, not only will axial symmetry obtain, but the final occluded star will be characterized by three and only three parameters, namely M, J, and Q. This kind of result represents only one of the directions in which "black hole physics" has been developed to a rather striking extent. Again, there is "black hole thermodynamics" in which an entropy S_b and an absolute temperature T_b — both certain specific functions of M, J, and Q — are ascribed to a Kerr-Newmann black hole in such a way that certain relations governing the behavior of black holes become *formally* identical with the basic relations and inequalities of phenomenological thermodynamics. So far this state of affairs is merely analogical. It becomes otherwise *if* one is prepared to accept that a black hole after all radiates energy, particulate and otherwise, its spectrum being the same as that of thermal omission from a traditional black body at temperature T_b. The quantum mechanical arguments which lead to that result, however, rest on rather shaky foundations, nor is it always easy to say whether the ideas and assumptions underlying classical thermodynamics are safeguarded. Be that as it may, to present black hole physics in detail would involve us in arguments of great length and much subtlety, both mathematical and physical. Therefore—and perhaps also because the actual existence of occluded stars has not yet been definitively established—we may fairly take the view that to go further would lead us outside the scope of these lectures.

We might now continue our search for further solutions of physical interest, though as I have already remarked, as a matter of practical realities we would have to confine ourselves to vacuum solutions. By the same token success is likely to be elusive unless the metric is of a special kind. This will be the case when, on the one hand, it has symmetries, that is, admits a p-parameter group of motions, $0 < p \leq 10$—the metric has p Killing vectors—and, on the other, when the Riemann tensor has special algebraic properties. By systematizing these the possibility of constructing useful classifications of vacuum fields lies at hand. This can be done in various more or less equivalent ways. It is, however, beyond our purpose to develop the somewhat formidable mathematical machinery—such as the spinor calculus, for example—which would make it possible to carry through this programme elegantly and in detail. Yet, so that these general remarks might not be altogether too nebulous, let me give a very abbreviated survey of one particular scheme of classification. Behind it lies a calculus based on the isomorphism between the proper homogeneous Lorentz group and the complex three-dimensional rotation group which legitimizes what we are about to do; but we need not concern ourselves with it explicitly and use more rudimentary, if much less elegant, means instead.

In place of the Riemann tensor we shall deal with the tensor

$$C_{ijkl} := R_{ijkl} + g_{i[k}R_{l]j} - g_{j[k}R_{l]i} - \tfrac{1}{3}g_{i[k}g_{l]j}R. \qquad (13.22)$$

It is known as the Weyl tensor. It has the same symmetries as R_{ijkl} but has the convenient property of being entirely trace-free. In passing, the metric is conformally Euclidean if and only if $C_{ijkl} = 0$, that is to say, by transformation of coordinates it can be brought into the form $ds^2 = \Phi \eta_{ij} dx^i dx^j$, where Φ is a scalar function. Since $R_{ij} = 0$ here, the Riemann and Weyl tensors coincide, but the classification of C_{ijkl} retains its usefulness even when $R_{ij} \neq 0$, since its algebraic properties are not affected. C_{ijkl} is in effect what remains when R_{mn} is "taken out of" R_{ijkl} and it has ten linearly independent components.

Now, since C_{ijkl} is skew in its first pair and its second pair of indices, each can in effect take only six combinations of values, namely $23, 31, 12, 14, 24, 34$. Introduce collective indices $A, B, \ldots = 1, 2, 3$ and $A', B', \ldots = 1', 2', 3'$ according to the scheme $23, 31, 12 \leftrightarrow 1, 2, 3$ and $14, 24, 34 \leftrightarrow 1', 2', 3'$. Then C_{ijkl} splits up into three 3×3 matrices $\mathbf{M} := C_{AB}, \mathbf{N} := C_{AB'}, \mathbf{L} := C_{A'B'}$. All three are symmetric on account of the symmetry $C_{ijkl} = C_{klij}$. The further condition $C_{i[jkl]} = 0$ implies that $\mathbf{N}$ is trace-free. There remains the condition $g^{il}C_{ijkl} = 0$. Take coordinates such that $g_{ij} = \eta_{ij}$ at the point considered and choose first $j, k = 23, 31, 12$. One finds that $\mathbf{M} = -\mathbf{L}$. Next, $j, k = 11, 22, 33$ gives the equation $\operatorname{tr}\mathbf{M} = C_{AA} + C_{A'A'}$, where tr denotes the trace and $A'=1'$ when $A=1$, and so on. Addition of these three equations gives $3\,\mathrm{tr}\,\mathbf{M}=\mathrm{tr}\,\mathbf{L}+\mathrm{tr}\,\mathbf{M}=0$ since $\mathbf{L}+\mathbf{M}=0$. Hence $\mathbf{L}$ and $\mathbf{M}$ are also trace-free. The remaining choices of j, k give nothing new. We are thus left with the two trace-free symmetric matrices $\mathbf{M}$ and $\mathbf{N}$. They may be combined into one complex symmetric trace-free matrix $\mathbf{P}=\mathbf{M}+i\mathbf{N}$.

From here on one can proceed in various equivalent ways. Perhaps most simply one considers the eigenvectors and eigenvalues of $\mathbf{P}$. Since $\mathbf{P}$ is trace-free the sum of its eigenvalues vanishes. Write α and β for the number of its distinct eigenvectors and eigenvalues, respectively. $\beta=1$ of course implies that all eigenvalues are zero. Then C_{ijkl} is characterized invariantly by its so-called Petrov type as follows. In increasing order of specialization, type I: $\alpha=3, \beta=3$; type D: $\alpha=3, \beta=2$; type II: $\alpha=2, \beta=2$; type N: $\alpha=2, \beta=1$; type III: $\alpha=1, \beta=1$. Type I is called algebraically general, all others algebraically special, or degenerate. Type O, which has $\mathbf{P}=0$, has been left out of account since a vacuum metric with $C_{ijkl}=0$ is Euclidean. A different procedure for arriving at the Petrov classification leads to the conclusion that a metric is degenerate if and only if there exists a null vector k_s such that $k_{[i}C_{j]kl[m}k_{n]}k^k k^l$. In particular, it is of type N if $C_{ijkl}k^l=0$, and equation which implies that $C_{ij[kl}k_{m]}=0$. This bears a strong formal resemblance to the equations which define the class of null Maxwell fields, namely, $f_{ij}k^j=0$, $f_{[ij}k_{l]}=0$.

Degeneracy of a metric entails the possibility of simplifying the explicit form of the field equations to be solved. To do justice to this subject, however, one cannot evade the use of sophisticated methods. The time has therefore come for us to abandon it and merely take a glance at what is perhaps the simplest kind of type N nonstatic vacuum metrics to which these methods lead. Generically, with $x:=x^1$, $y:=x^2$, $z:=x^3$, $t:=x^4$, $u:=t-z$, $v:=t+z$,

$$ds^2 = -e^{2(\alpha+\beta)}dx^2 - e^{2(\alpha-\beta)}dy^2 + du\,dv, \qquad (13.23)$$

where α and β are functions of u alone. Then, with $A:=\alpha''+\alpha'^2+\beta'^2$, $B:=\beta''+2\alpha'\beta'$, it follows in an obvious notation that

$$R^x_{\ uux}=A+B, \qquad R^y_{\ uuy}=A-B, \qquad R_{uu}=2A \qquad (13.24)$$

are the only distinct nonvanishing components of the Riemann and Ricci tensors, primes indicating derivatives with respect to u. Hence the field equations are satisfied if only $\alpha''+\alpha'^2+\beta'^2=0$ and the metric will then not be Euclidean provided $\beta'e^{2\alpha}$ is not constant. (13.23) can be transformed in an interesting and useful way when it satisfies the field equations, that is,

$A=0$. Set

$$x=e^{-\alpha-\beta}X, \qquad y=e^{-\alpha+\beta}Y, \qquad u=U,$$

$$v=V-(\alpha'+\beta')X^2-(\alpha'-\beta')Y^2. \qquad (13.25)$$

Then (13.23) becomes

$$ds^2=-dX^2-dY^2+dU\,dV-(X^2-Y^2)F(U)\,dU^2, \qquad (13.26)$$

where $F(U):=\beta''+2\alpha'\beta'$. This, then, is a solution of $R_{ij}=0$, where F is now an arbitrary function of U. It is manifestly Euclidean when $F(U)$ vanishes, consistently with (13.24).

The metric (13.23) cannot depend periodically on u since the vanishing of A implies that α'' has a fixed sign. On the other hand one can arrange α and β to vanish outside the region $0 \le u \le a$, say, where a is some constant. Then (13.23) represents a spacetime in which two Euclidean regions I and III are separated by a non-Euclidean region II. In the language of temporal evolution, the boundaries of II, that is, $z=t$ and $z=t-a$ are continuously displaced with unit speed. In region II one may simply choose one of the functions α, β arbitrarily and calculate the other from the equation $A=0$. In regions I and III, that is, when $u \le 0$ and $u \ge a$, respectively, one has $A=B=0$, of course, or $(e^{\alpha+\beta})''=0$ and $(e^{\alpha-\beta})''=0$. $e^{\alpha+\beta}$ and $e^{\alpha-\beta}$ are therefore both linear functions of u. To complete the solution it then remains to ensure that the junction conditions are satisfied, that is to say, that α, β and their first derivatives are continuous at $u=0$ and $u=a$. This may entail the appearance for singularities in I or III or both, but they must be spurious for we already know that transformations of the kind (13.25) lead to the regular Euclidean metric $ds^2=-dX^2-dY^2+dU\,dV$.

Now, since (13.23) has $g_{i4}=\delta_{i4}$ and therefore $\Gamma^i_{44}=0$, the geodesic equations are satisfied by $x^a=$ constant, $x^4=s+$ constant. For the sake of simplicity, consider a solution such that in the region I $\alpha=\beta=0$. Two neighboring test particles P and Q with fixed coordinates x^a, $x^a+\delta x^a$ are then at rest in region I as long as $t \le z$, granted that $\delta z \ge 0$. When $z+\delta z < t < z+a$, P and Q are within II. Here their mutual distance δl is given by (10.2) which is meaningful even for nonstatic metrics as long as one does not attempt to integrate dl. Then

$$(\delta l_{\mathrm{II}})^2=e^{e(\alpha+\beta)}(\delta x)^2+e^{2(\alpha-\beta)}(\delta y)^2+(\delta z)^2. \qquad (13.27)$$

Since $ds=dt$ the relative speed of P and Q is therefore

$$v_{\mathrm{II}}=\left[(\alpha'+\beta')e^{2(\alpha+\beta)}(\delta x)^2+(\alpha'-\beta')e^{2(\alpha-\beta)}(\delta y)^2\right]/\delta l_{\mathrm{II}}. \qquad (13.28)$$

If the displacement from P to Q is longitudinal, that is, in the direction of z alone increasing, $v=0$, whereas when it has a transverse component v will in general fail to vanish.

There is motion where there was none before. We should be able to extract useful work by allowing friction to bring P and Q to relative rest. Certainly we seem to have a pretty good indication that a particle system can acquire kinetic energy as the result of its encountering a radiative pulse. Radiative pulse? Why radiative? At this stage I can offer no justification for the use of this terminology other than formal analogies with plane wave solutions of Maxwell's vacuum equations. They represent null fields and the analogy of these with Petrov type N solutions I have already remarked upon. In both cases the "wavefronts"—hypersurfaces on which second derivatives, of A_i in the one case and of g_{ij} in the other, are discontinuous—propagate with unit speed; and so on. Sometimes one is tempted to speak of a radiative solution when the g_{ij} depends for example on $x^3 - x^4$ alone, where x^3 is a spacelike and x^4 a timelike coordinate; but this property of the metric is easily destroyed by a coordinate transformation. Without coordinate-independent criteria it isn't sensible to characterise some particular solution as being radiative.

However that may be, the conclusion that useful work might be extracted from P and Q is dubious for the following reason. P and Q are test particles and therefore "have no mass"—a phrase to be interpreted as meaning that their masses are so small that the approximation $T^{ij}=0$ is to all intents and purposes justified. One cannot have it both ways: in that case their kinetic energies which we propose to convert into heat must also be approximated by zero. Secure conclusions could only be reached on the basis of solutions physically far less hopelessly unrealistic than (13.23); but in practice they seem to be unattainable. One's first thought might therefore be to take refuge in the linear approximation. Before I raise objections to this—to some extent I have already discussed them—we shall take a brief look at this at the beginning of the next lecture.

LECTURE 14

According to equation (8.7) the linearized vacuum metrics must satisfy the equation $\Box l_{ij} = 0$, where $g_{ij} = \eta_{ij} + h_{ij}$, $l_{ij} := h_{ij} - \frac{1}{2}\eta_{ij}h$ and the coordinate conditions $l^{ij}{}_{,j} = 0$ must be satisfied. Formally one has the equation of a source-free field h_{ij} in a Euclidean spacetime. One may therefore use colloquial language without much risk of confusion. In particular one has monochromatic plane-wave solutions

$$h_{ij} = a_{ij}\exp\left(ik_l x^l\right) =: a_{ij}e^{iS}, \tag{14.1}$$

the a_{ij} being constants. The physical field is the real part of this—recall the locally plane-wave Maxwell vacuum fields we considered in the ninth lecture. The wave equation and coordinate conditions require that k_i be a null vector and that $a_{ij}k^j = \frac{1}{2}k_i a_j{}^j$, respectively. Therefore it seems that of the ten amplitudes a_{ij} six are mutually independent. We must be careful here, for some particular solution of the form (14.1) may be spurious to the extent that it corresponds to a vanishing Riemann tensor: l_{ij} has appeared as the result of a mere infinitesimal coordinate transformation $x^i \to x^i + \xi^i$. A transformation of this kind changes, as we know, h_{ij} into $h_{ij} := h_{ij} + 2\xi_{(i,j)}$. These will still satisfy the coordinate conditions provided $\Box \xi_j = 0$. Choose $\xi_j = -i\epsilon_j e^{iS}$, with constant ϵ_j. In effect a_{ij} changes into $\hat{a}_{ij} := a_{ij} + 2\epsilon_{(i}k_{j)}$.

By means of the coordinate conditions four of the a_{ij} may now be expressed in terms of the other six. These six components may then be simplified by suitably choosing the four constants ϵ_j. As a result only two independent constants remain. For example, for a wave progressing in the direction of the x^3-axis — $k_1 = k_2 = 0$, $k_3 = -k_4$ — one ends up with $\hat{a}_{ij}$ having only $\hat{a}_{11} = -\hat{a}_{22}$, $\hat{a}_{12} = \hat{a}_{21}$ as nonvanishing components. In this sense, then, a plane wave has only two degrees of freedom. In other words, there are just two independent states of polarization and the "polarization tensor" $\hat{a}_{ab}$ is trace-free. The equation corresponding to (13.27) for the relative displacement δl of two test particles P and Q located transversally

125

to the direction of propagation is

$$\delta l = \rho\left(1 + \tfrac{1}{2}h_{11}\cos 2\psi + \tfrac{1}{2}h_{12}\sin 2\psi\right),$$

where $\delta x^1 := \rho\cos\psi$, $\delta x^2 =: \rho\sin\psi$. A set of particles Q lying at some time on a circle with P as center will therefore at other times generally lie on an ellipse with P as center, so that one has a typical tidal effect. Superposition of monochromatic waves with different angular frequencies k_4 will give a general plane wave progressing in the x^3-direction.

We could go on to investigate more general solutions with wavelike character when $T_{ij} \neq 0$. Alternatively, one might suppose that the actual metric $\hat{g}_{ij}$ differs from a given metric g_{ij} only to such an extent that terms depending nonlinearly on $h_{ij} := \hat{g}_{ij} - g_{ij}$ may be disregarded. Let all operations here refer to g_{ij}, not $\hat{g}_{ij}$. Then by an infinitesimal transformation of coordinates one can always arrange that $l^j_{i\;;j} = 0$, where $l_{ij} := h_{ij} - \tfrac{1}{2}g_{ij}g^{kl}h_{kl}$. Then, if g_{ij} itself satisfies the vacuum equations, the $\hat{g}_{ij}$ will do likewise if

$$\Box\, l_{ij} - 2R_{imnj}l^{mn} = 0. \tag{14.2}$$

As in the case of the Maxwell field, one can study the counterpart to the geometric-optical limit by setting

$$l_{ij} = \operatorname{Re}\left(a_{ij}e^{i\eta S}\right) \tag{14.3}$$

where the spacetime variation of the amplitudes a_{ij} is again supposed to be sufficiently small compared with that of S. One thus regains the equation

$$g^{ij}S_{,i}S_{,j} = 0. \tag{14.4}$$

The "disturbance" h_{ij} therefore propagates with unit speed. (14.4) is the so-called characteristic equation for equation (14.2), that is, for the propagation of disturbances governed by this. The l_{ij} and their first derivatives are required to be continuous—one wants no discontinuities in the Γ^k_{ij} which enter into the geodesic equations. Recall that a wavefront, or "characteristic hypersurface," is defined by the presence of discontinuities of the second derivatives across it. Let $S(x) = 0$ be the equation of some hypersurface. Proceeding somewhat heuristically, set $l_{ij} = \psi_{ij}(x)S^2\theta(S)$ in (14.2), where θ is a step function. The ψ_{ij}, together with their first and second derivatives, are supposed to be continuous. Virtually by inspection, there is only one term on the left of (14.2) which will not vanish as a consequence of $S(x) = 0$. Its absence requires just that (14.4) be satisfied.

As a digression, let us take note of the important fact that (14.4) is the characteristic equation even in the absence of any approximations. To verify this claim, recall (8.6) and write out R_{ij} as follows:

$$R_{ij} = \tfrac{1}{2} g^{mn} g_{ij,mn} - g_{(i,j)} + N_{ij},$$

where $g_i := g^{mn}(g_{im,n} - \tfrac{1}{2} g_{mn,i})$. In the interests of simplicity we now impose the four coordinate conditions $g_i = 0$. The equation $R_{ij} = 0$ then becomes

$$g^{44} g_{ij,44} + U_{ij} = 0, \tag{14.5}$$

where U_{ij} contains no second derivatives with respect to x^4. The initial value problem—or "Cauchy problem"—is now this: to find the functions g_{ij}, given their values and those of the derivatives $g_{ij,4}$ on some hypersurface C which has the equation $S(x) = 0$. In some neighborhood of this surface one may seek to accomplish this by evaluating successive derivatives of the g_{ij}. Note incidentally that the $g_{ij,4}$ cannot be arbitrarily prescribed, since the coordinate conditions impose constraints on them. Suppose first that $S(x)$ has been arranged to have the specific form $S(x) = x^4$. All the $g_{ij,ab}$ and $g_{ij,a4}$ are found by direct differentiation. The $g_{ij,44}$, however, must be obtained from (14.5). This is impossible if and only if g^{44} vanishes on C. Now go over to arbitrary coordinates—the coordinate conditions need no longer be satisfied. $g^{44} = 0$ becomes on C

$$x^4{}_{,i'} x^4{}_{,j'} g^{i'j'} = g^{i'j'} S_{,i'} S_{,j'} = 0.$$

The Cauchy problem is thus insoluble if C satisfies the characteristic equation which means that discontinuities of the second derivatives occur on and only on characteristic hypersurfaces. Since Maxwellian characteristic hypersurfaces obey the same equation one concludes quite generally that discontinuities of the Maxwell tensor and those of the metric tensor propagate with unit speed, as I already stated at the end of the preceding lecture.

Finally one can find rays. A ray cuts a family of characteristic hypersurfaces orthogonally. Then if the equation of a ray is $x^i = x^i(\lambda)$, where λ is a suitable parameter, $dx^i/d\lambda = S^{,i}$. On the other hand, by differentiation of (14.4), $S^{,i} S_{,i}{}^{,j} = S^{,i} S_{,j}{}^{,i} = 0$. It follows that $d^2 x^i/d\lambda^2 + \Gamma^i{}_{jk} (dx^j/d\lambda)(dx^k/d\lambda) = 0$. In other words, the rays are null geodesics. This general result is in harmony with the conclusion which we reached in the special context of plane-wave solutions.

Returning to (14.1)–(14.4), let us not lose sight of the fact that these are part of a linearized problem—that was the object of the exercise. Yet, by linearizing much of the physics is lost. Furthermore, is the result of linearization a first step in a converging sequence of approximations? This question I have already discussed at some length and I shall not do so again. We do not know what weight to attach to conclusions drawn from the linearized equations. Perhaps a cogent objection to them is the following. Suppose h_{ij} has been calculated. Now substitute $\eta_{ij} + h_{ij}$ in the field equations, but no longer reject any terms. As a result we end up with some formal energy tensor $\tilde{T}_{ij}$. I say formal, because it may be of a kind which cannot possibly represent some realistic physical system—for instance $\tilde{T}_4^{\ 4}$ might turn out to be negative. However, even when such difficulties are ignored, the fact is that we will have a metric describing physical circumstances different from those which we set out to describe. In particular, if T_{ij} was intended to be zero, it will usually not be zero. In short, the linear approximation should be thought of as approximating the exact solution of the equations $G_{ij} = \kappa \tilde{T}_{ij}$, rather than that of the equation $G_{ij} = \kappa T_{ij}$, where T_{ij} was the energy tensor given originally.

In view of all this the linearized equations—particularly the analogies which exist between their solutions and those of the usual Maxwell equations—should principally be regarded as merely a heuristic guide. On this basis one can develop criteria intended to discriminate between those exact vacuum metrics which are to be regarded as being radiative and those which are not. Such criteria are mostly related to the Petrov classification, as I indicated before. A good many alternatives have been proposed, some of which either overlap or are mutually equivalent; or else they fail to be exhaustive or may be deficient in some other way. To pursue this theme is, however, quite beyond the scope of these lectures.

Again, guided by the linear approximation one may envisage the vacuum region far from some material nonstatic system. Granted that in this remote region the metric is nearly Euclidean one may under appropriate circumstances attempt an asymptotic expansion of the Riemann tensor in inverse powers of some suitable coordinate r which increases with distance. For certain types of metrics it turns out that $R_{ijkl} = \Sigma_{n=1} W^{(n)}_{\ ijkl} r^{-n}$, where $W^{(n)}_{\ ijkl}$ has all the algebraic properties of the Weyl tensor and is of Petrov type N, III, II, I for $n = 1, 2, 3, 4$, respectively. One can then speak of a radiation zone or far field for $n = 1$; a semifar field for $n = 2$; and so on.

Interesting and suggestive though all such considerations undoubtedly are, the fact remains that they are formal and rest, moreover, on an analogy which may, after all, be totally misleading. When one talks about a wave zone here one presumably has the idea of the electromagnetic wave zone at the back of one's mind. The retarded Maxwell field associated with some finite source is radiative if it falls off asymptotically as the inverse distance from the source, so that there is then a finite flux of energy across a remote closed surface surrounding the source, at the expense of its mechanical energy. At last we are coming to the heart of the matter: surely whenever we interest ourselves in radiation we are contemplating the noninductive transfer of energy. That is, I believe, the context in which one must ultimately think about the division of metrics into radiative and nonradiative classes.

By way of illustration contemplate a metric representing a state of affairs which perforce I shall have to describe in colloquial terms. There are two like, well-separated adiabatically isolated systems E and A. Each consists of two small bodies mutually connected by an imperfectly elastic spring. That of E is extended beyond its equilibrium length by some constraint, while A is unconstrained and in equilibrium. At some time the constraint collapses, so that E will temporarily vibrate. By mechanical and calorimetric measurements one can find the overall change ϵ of the energy of E. Just when a test light pulse emitted by E when it began to vibrate reaches A, the latter begins to vibrate, and this vibration will also die out. The change α in the energy of A is necessarily positive. If $\epsilon < 0$ and $|\epsilon| > \alpha$ one will not have any hesitation to say that energy has been transferred radiatively from E to A and therefore that the metric is radiative. Of course, to find an exact metric representing a physical state of affairs as enormously complicated as this is quite out of the question. Still, we must suppose that it exists even if we cannot find it and physical principles are after all our main concern, not mathematical manipulations. Notice that the use of the descriptive phrase "energy has been transferred from E to A" indicates implicit causal notions; and we all know of the difficulties associated with these. At any rate, we tend to go beyond the mere facts established by observation by thinking of energy moving from E to A in a way capable of detailed temporal description.

Here we have come back full circle to the lengthy discussion which occupied the early part of the ninth lecture. However, now it seems that I can no longer evade talking about energy in the sense in which one does so in traditional, or, if you like, local field theories. Or can I? In the light of all the difficulties which we encountered previously, might it not be best to be quite hardheaded and to abandon one's traditional ideas of energy, its localization, its conservation, and the like? We invent a particular physical quantity because it is useful in the formalization of observational results, in the creation of theories which reflect regularities of the world. It is useful, that is to say, because under circumstances originally envisaged it conforms to certain fixed rules. When it ceases to do so under more general circumstances—for instance, when it then fails to satisfy some law of conservation which is satisfied originally—we try to modify it; but when no acceptable modification presents itself we must abandon it.

What do we find when we examine energy from this point of view? Energy was originally defined in a mechanical context since the change of the energy of a system, that is of its energy of motion, could be accounted for by the work visibly done on it. When no work is done the energy is fixed — it is conserved. As soon as it was realized that a system may acquire kinetic energy merely by changing its position, conservation was rescued by inventing potential energy. However, even that is not good enough: a fluid upon being stirred acquires energy of motion, but eventually this apparently just disappears. Of course, we are now talking about thermodynamic systems and the law of conservation of energy is rescued this time with the aid of the idea of adiabatic isolation. Finally, a body initially at rest will, in general, finally be moving as the result of the action of a pulse of electromagnetic radiation. This time it turns out that the law of conservation of energy can still be maintained because— as a consequence of the structure of Maxwell's theory— one can consistently ascribe a certain energy density and energy flux to the electromagnetic field itself. This familiar conclusion is contained in the existence of the symmetric Maxwell energy tensor $T_M{}^{ij}$ which, according to equation (7.18) is such that $T_M{}^{ij}{}_{,j} = f^{ij} j_j$. Here we have merely a partial derivative because the classical Maxwell theory operates within a Euclidean framework and the coordinates are Cartesian. Integrating over a three-dimensional region with boundary B which contains all charges present, it follows by an application of Gauss' theorem that the appearance of mechanical energy and momentum within B is exactly balanced by corresponding fluxes across B, and if the field vanishes on B the total energy and momentum within B are conserved. There is one bit of murkiness here in as far as $T_M{}^{ij}$ is not uniquely defined: one could add to it an arbitrary divergence of the form $U^{i[jk]}{}_{,k}$. However, we adhere to T^{ij} by prescription. Then certainly the energy density $T^{44} > 0$ unless the field vanishes, and this inequality is unaffected by the choice of coordinate system.

Once again it has proved possible to satisfy the demand for the maintenance of the law of energy conservation. Here I am saying that "a demand has been satisfied" for in the course of time the existence of a conserved energy seems almost to have acquired the status of a metalaw. As regards fields in interaction with matter it would go something like this: as a consequence of the equations of any admissible field theory it must be possible to define a field energy density such that an isolated system consisting of matter and field possesses a conserved energy. Furthermore, the field energy density is usually required to be positive. One can easily see why one might wish to have such a metalaw— let us call it *P*4. Overall energy conservation would be assured and the presence of the field would not lead to perpetual motion of the first kind; nor would quantum theory have to survive without Hamiltonians. *P*4 is in fact accommodated automatically by subjecting fields to *P*2 and by requiring them to be Lorentz covariant—their equations of motion must be generated by a variational principle the Lagrangian of which is invariant under inhomogeneous Lorentz transformations. Then as a consequence of invariance under spacetime translations Nöther's theorem implies the existence of four continuity equations $T^{ij}{}_{,j}=0$.

Before concluding that this is a satisfactory state of affairs we must however reflect that as we increase the number of metalaws into our general scheme we run an increasing risk of finding some of them to be in mutual conflict. In fact, that is exactly the situation with which we are now confronted, for *P*1 and *P*4 are in general not compatible with each other. *P*1 and *P*4 both demand proper form invariance: *P*4 does so when only the field ψ in question is a dynamical object whilst the metric is by prescription an absolute[II] object, $g_{ij}=\eta_{ij}$, whereas *P*1 does so when the field and the metric are both dynamical objects. In the latter case the equations for ψ will not be properly form invariant under Lorentz transformations unless the dynamical object g_{ij} happens to go into itself, or in other words, unless the solution of the equations it has to satisfy happens to have some special form. In short, unless g_{ij} happens to admit one or more Killing vectors, all we have available is the general consequence of equations (8.1) that $T^{ij}{}_{;j}=0$; but these are not continuity equations in the usual sense of that term. To be quite explicit, it is useful to write it in a slightly different form. To this end, if f is any quantity—any indices it may have have been suppressed—let the replacement of the kernel symbol f by its German counterpart indicate multiplication by $(-g)^{1/2}$: $\mathfrak{f}=(-g)^{1/2}f$. Then $T^{ij}{}_{;j}=0$ takes the equivalent form

$$\mathfrak{T}_i{}^j{}_{,j}=\tfrac{1}{2}g_{jk,i}\mathfrak{T}^{jk}. \tag{14.6}$$

On account of the presence of the term on the right no conservation laws will result by integration.

It seems we must give up either *P*1 or *P*4. We are unwilling to abandon *P*1, so that it is *P*4 which has to go. It may be a bitter pill to swallow, but swallow it we must, at any rate as long as we do not recognize the superenergy tensors mentioned earlier as appropriate analogs of the usual objects of traditional field theory. Still, as I stressed before, we have no physical field here—the g_{ij}-field is no "field of force"—the situation is unfamiliar, and insistence on *P*4 the result of habit.

We appear to be confronted with a dilemma, for we seem to have no means of analyzing, for example, the way in which part of the initial energy of some system E subsequently comes to be part of that of some other system A. Faced with this, the best one can do is to effect a compromise. It goes something like this. First one rewrites (14.6) as a formal equation of continuity $\mathfrak{S}_i{}^j{}_{,j}=0$. This can always be achieved—in fact $\mathfrak{S}_i{}^j$ is not uniquely determined. If $t_i{}^j:=\mathfrak{S}_i{}^j-\mathfrak{T}_i{}^j$, one then usually speaks of $t_i{}^j$ as an "energy-momentum complex," or energy complex, for short. Sometimes it is also referred to as a "pseudotensor" presumably on the grounds that it is made up of the components of the metric tensor and its derivatives alone. Also, it turns out to be a tensor under linear transformations of coordinates. It is here that the real compromise begins, since one now has to impose coordinate conditions and asymptotic conditions. Suppose $T_i{}^j$ vanishes outside some three-dimensional spacelike region. Then require that for a suitable choice of coordinate system x^j the following conditions are satisfied:

$$g_{ij}=\eta_{ij}+O(r^{-1}), \qquad g_{ij,k}=O(r^{-2}) \qquad (14.7)$$

as $r:=(\delta_{ab}x^ax^b)^{1/2}\to\infty$. Integration over all values of the x^a then leads to the conclusion that the four quantities

$$P_i:=\int\mathfrak{S}_i{}^4 d^3x \qquad (14.8)$$

are globally conserved. P_i is finite, transforms as a covariant vector under linear transformations, and is unaffected by transformations of coordinates which preserve (14.7) and reduce to the identity for $r\to\infty$.

Before going on let me just show how $t_i{}^j$ may be constructed, using only the most elementary methods. Recall that the addition of any ordinary divergence to a Lagrangian leaves its functional derivative unaffected. Therefore if one adds $2(\mathfrak{g}^{i[j}\Gamma^{k]}{}_{ij})_{,k}$ to $\mathfrak{R}$ one gets the equivalent Lagrangian

$$\mathfrak{R}^*=2\mathfrak{g}^{ij}\Gamma^m{}_{i[j}\Gamma^k{}_{k]m}, \qquad (14.9)$$

from which second derivatives of the g_{ij} are absent. Since $G^{ij}=\delta R^*/\delta g_{ij}$, G^{ij} is therefore of the second order, as we already know. Now write $\mathfrak{r}^{ij}:=\partial\mathfrak{R}^*/\partial g_{ij}$, $\mathfrak{r}^{ijk}:=\partial\mathfrak{R}^*/\partial g_{ij,k}$ and define the quantity

$$\mathfrak{c}_i{}^j:=g_{mn,i}\mathfrak{r}^{mnj}-\delta^j{}_i\mathfrak{R}^*, \qquad (14.10)$$

reminiscent of the canonical energy tensor of conventional field theory. By differentiation

$$c_{i\,,j}^{\,j}=g_{mn,i}\left(\mathfrak{r}^{mnj}{}_{,j}-\mathfrak{r}^{mn}\right)=-g_{mn,i}\delta R^*/\delta g_{mn}.$$

In view of the field equations the expression on the right is just $-\kappa g_{mn,i}\mathfrak{X}^{mn}$. Upon comparing this with (14.6), it follows that

$$\mathfrak{t}_i^{\,j}:=(2\kappa)^{-1}\left(\delta_i^{\,j}\mathfrak{R}^*-g_{mn,i}\mathfrak{r}^{mnj}\right) \tag{14.11}$$

is an object of the required kind, namely the so-called Einstein energy complex. $\mathfrak{t}^{ij}$ is not symmetric, so that is gives rise to no sensible formal angular momentum complex. Alternative symmetric complexes have therefore been constructed, but I shall leave these aside.

Reverting to the formal continuity equation $\mathfrak{S}_{i,j}^{\,j}=0$, one might *say*— granted (14.7)—that spacetime in itself has a total energy and momentum $p_i:=\int \mathfrak{t}_i^{\,4}\,d^3x$, to "account for" the global conservation of P_i. This is acceptable as long as one does not impute to p_i more than an accounting function. To ascribe to the "energy and momentum of spacetime" a more conventional meaning hardly makes sense. Bear in mind that p_i is not even defined unless at least the first member of (14.7) obtains; but who is to say without the warrant of observational results what the character of the metric will in fact be when r becomes large? At any rate, granted our assumptions, for a static sphere of the kind we considered earlier one finds that $P_4=M$, $P_a=0$. Here we might just note that $m^*:=\int \mathfrak{X}_4^{\,4}\,d^3x$ is not what we called the inertial mass m. Taking the metric in the generic form (12.36) and writing $dV:=|\det g_{ab}|^{1/2}\,d^3x$ for the element of proper volume, $m=\int\rho\,dV$ whereas $m^*=\int\rho\sqrt{g_{44}}dV$. Moreover, $M<m^*$, so that $\int \mathfrak{t}_4^{\,4}d^3x$ is negative.

For nonstatic systems there is no reason to take it for granted that the second of the conditions (14.7) will be satisfied. On the contrary, whether by analogy with electrodynamics or on the basis of the linear approximation, in regions remote from some nonstatic system one would expect that $g_{ij,k}=O(r^{-1})$ rather than $O(r^{-2})$. The convergence of the integral then becomes problematic. If one imposes suitable coordinate conditions and furthermore supposes that the system was quiescent for an infinite time in the past, one can still give reasonable prescriptions for calculating the overall energy and its rate of change. It is clear, however, that more and more uncertainties appear.

By tradition one has not been satisfied with global conservation laws for fields, but has tended to reduce them to local terms by the ascription to the field of local densities of energy, momentum and their fluxes. One should, however, not expect that such a reduction from the global to the local

134

standpoint will necessarily have more than formal significance. How, otherwise, is one to understand the Poynting flux in crossed electrostatic and magnetostatic fields? Think, too, of entropy production in process thermodynamics. The global rate of entropy production in a system is a well-defined quantity; but locally it can be changed willy-nilly by a thermal gauge transformation. As far as $t_i{}^j$ is concerned, its significance is even more obscure for, as I have stressed continually, there is no field. Consistently with this, $t_i{}^j$ can be reduced to zero at any given event, simply by choosing normal coordinates there. In short, we take the $t_i{}^j$ to fulfil merely a bookkeeping function. We refuse to accord to them the status of physical entities since, viewed locally, they lack the minimal degree of permanence which one normally demands of a physical entity. Even for accounting purposes the acceptability of the local quantities $t_i{}^j$ hinges on the satisfaction of appropriate coordinate and boundary conditions—and these are nonlocal. Here a reference to the particular plane-wave metric (13.26) which violates them is very instructive. For this all the components of $t_i{}^j$ vanish, which seems odd, bearing in mind our previous discussion concerning the motion of test particles, though admittedly the doubts about this expressed at the end of the thirteenth lecture must make us wary of drawing rash conclusions.

Undeniably we have made no clear-cut headway toward answering the questions raised in the ninth lecture. The implications of physically realistic radiative solutions remain in doubt, if for no other reason than that no such solutions are available and approximations are beset with such perplexities as to leave basic questions unresolved. Yet for good reasons we want to inquire specifically into the behavior of a binary star system for it represents a quite unambiguous test case. Qualitatively observation is confronted with a simple trichotomy—on the average the mutual distance between the components either increases, remains constant, or decreases. On the basis of astronomical evidence, especially if it manages to be quantitative, one will then be able to decide whether essentially heuristic theoretical methods have led to correct predictions. If they have not there is no sense in trying to place them on more secure foundations.

LECTURE 15

At the end of the preceding lecture we decided to carry through some kind of analysis of the behavior of a double star system. Of necessity we shall have to be content with piling approximation upon approximation. In particular, we shall have to throw all our reservations to the winds and operate throughout on the level of the linear approximation, even if we have to ignore some inconsistencies which this involves: of these more later. Furthermore, where the energy complex $t_i{}^j$ plays its part in the calculations, we simply take it for granted that it is playing its proper role. In this way one at least arrives at some explicit formulae to be tested directly against the evidence of observational results: and that is, after all, the object of the exercise.

Suppose, then, that M is some bounded nonstatic material system. Straight away we assume that even within it the linear approximation is adequate. Of course, one says this with tongue in cheek: in the strictly linear approximation the field equations place no limitations on the motions of bodies: the earth would move as if the sun did not exist. Further, if linearization is to be justified, $|T^{ia}|$ must be sufficiently small compared with T^{44}, that is to say, internal motions and stresses must be small. Now, in the linearized domain all calculations proceed formally as if one were dealing with fields defined on a Euclidean spacetime. One may therefore conveniently use a language and notation appropriate to this. In a manner which explains itself, (9.1), for instance, can be written as follows

$$l_{ij} = -(\kappa/2\pi) \int R^{-1} T_{ij}(\bar{\mathbf{r}}, t-R) \, d\bar{\mathbf{r}}, \qquad (15.1)$$

where $t := x^4$, $R^2 := |\mathbf{r} - \bar{\mathbf{r}}|^2$. We are taking it for granted that we should consider only the retarded solution, corresponding to radiation going out from M. At points remote from M the factor R^{-1} multiplying T_{ij} may be replaced by r^{-1}. On the other hand, since, as already assumed, the motions

135

are slow, retardation effects may be neglected within M, that is to say, $t-R$ may be approximated by $t-r$. If $f(\bar{\mathbf{r}}, t)$ is any function which depends on t, write $f^*:=f(\bar{\mathbf{r}}, t-r)$ and then (15.1) becomes

$$l_{ij} = -(\kappa/2\pi)r^{-1}\int T_{ij}^* \, d\bar{\mathbf{r}}. \tag{15.2}$$

Far from M the l_{ij} represent almost plane waves, going outward from M. We saw previously that given such a wave traveling in the direction of the x^3-axis, one could arrange by means of a suitable infinitesimal coordinate transformation $x^i \to x^i + \xi^i$ that h_{ij} had only the components $h_{11} = -h_{22}$ and $h_{12} = h_{21}$ which did not necessarily vanish. To free this result from the particular direction involved in it, one need only translate it into a form invariant under three-dimensional rotations; namely

$$h_{i4} = 0, \qquad h_a^{\ a} = 0. \tag{15.3}$$

As an incidental consequence of (15.3), since h now vanishes the distinction between h_{ij} and l_{ij} disappears. What can be achieved for a single wave can be achieved for any superposition of them by a suitable choice of ξ^i. In short, we can always adopt a coordinate system so that (15.3) holds. Consequently only the h_{ab} need now to be found.

First, from $T^{ij}_{\ \ ,j} = 0$ it follows trivially that $2T^{ab} = (\bar{x}^a \bar{x}^b T^{ij})_{,ij}$. Then

$$2\int T^{ab} \, d\bar{\mathbf{r}} = \int \left[(\bar{x}^a \bar{x}^b T^{ic})_{,i} + \bar{x}^a \bar{x}^b T^{c4}_{\ \ ,4} \right]_{,c} d\bar{\mathbf{r}} + \int \bar{x}^a \bar{x}^b T^{44}_{\ \ ,44} \, d\bar{\mathbf{r}}.$$

Let the integration extend over a region whose boundary wholly encloses M. By means of Gauss' theorem the first integral becomes a surface integral which vanishes since T^{*ij} vanishes on the boundary. What remains is just what we need in (15.2):

$$l^{ab} = h^{ab} = -(\kappa/4\pi)r^{-1}\int \bar{x}^a \bar{x}^b \ddot{\rho}^* \, d\bar{\mathbf{r}}, \tag{15.4}$$

dots denoting derivatives with respect to t. No account has yet been taken of the special choice of coordinate system which leads to (15.3). Clearly the transformation induced by ξ^i must be such as to remove the trace from h^{ab}, so that the latter finally becomes

$$h_{ab} = -(\kappa/4\pi)r^{-1}\ddot{Q}^*_{ab}, \tag{15.5}$$

where

$$Q^{ab} := \int \left(\bar{x}^a \bar{x}^b - \tfrac{1}{3} \delta^{ab} \bar{r}^2 \right) \rho \, d\bar{r} \tag{15.6}$$

is the so-called reduced quadrupole moment of M.

Considering all the while points remote from M, energy goes outward radially in all directions, at any rate in the formal bookkeeping sense which we accepted earlier. Over not too large regions each elementary wave is approximately plane. When computing $t_4{}^a$ we must include in the calculations only the two dynamical degrees of freedom which it possesses. In the case of a wave traveling in the z^3-direction, for example, we saw these to be described by the components $h'_{11} = -h'_{22}$ and $h'_{12} = h'_{21}$, with amplitudes $\hat{a}_{11}$, and so on. Here I have written h'_{ab} in place of h_{ab} to emphasize that they are transverse and trace-free components. To obtain $t_4{}^a$ from (14.11) is a little tedious. The result under the present simplifying circumstances is this:

$$t_4{}^a = (4\kappa)^{-1} h'_{bc,4} h'^{bc,a} = (4\kappa)^{-1} n_a \dot{h}'_{bc} \dot{h}'^{bc}, \tag{15.7}$$

for a wave in the direction of the unit three-vector $\mathbf{n}$. To avoid much trivial confusion, take all indices temporarily to refer to the metric $\delta_{ab} =$ diag $(1,1,1)$, so that the indices may all be written as subscripts, though the summation convention will be maintained. What we need to do now is to write down a covariant relation which expresses h'_{ab} in terms of h_{cd} and n_e in such a way that when $n_a = (0,0,1)$ only $h'_{11} = -h'_{12}$ and $h'_{12} = h'_{21}$ remain, for in that way we are extracting the required physically significant part of h_{ab}. The relation in question is

$$h'_{ab} = \left(P_{ac} P_{bd} - \tfrac{1}{2} P_{ab} P_{cd} \right) h_{cd}, \tag{15.8}$$

where $P_{ab} := \delta_{ab} - n_a n_b$ is a projection operator, so that $P_{ab} P_{ac} = P_{bc}$, $P_{ab} n_b = 0$, $P_{aa} = 2$. Upon inserting (15.8) into (15.7) one finds quite easily that

$$t_{4a} = (4\kappa)^{-1} n_a \left[\dot{h}_{bc} \dot{h}_{bc} - 2 n_b n_c \dot{h}_{bd} \dot{h}_{cd} + \tfrac{1}{2} \left(n_b n_c \dot{h}_{bc} \right)^2 \right], \tag{15.9}$$

bearing in mind that $h_{aa} = 0$. The total rate of radiation of energy is $r^2 \int t_{4a} n_a \, d\omega$, with $d\omega = \sin\theta \, d\theta \, d\phi$. Using (15.9) together with (15.5), the rate of change of the energy E of M is therefore

$$\dot{E} = -\kappa^{-1} \int \left(\dddot{Q}^*_{ab} \dddot{Q}^*_{ab} - 2 n_a n_b \dddot{Q}^*_{ac} \dddot{Q}^*_{bc} + \tfrac{1}{2} \left(n_a n_b \dddot{Q}^*_{ab} \right)^2 \right) d\omega.$$

Since $\int n_a n_b \, d\omega = (4\pi/3)\delta_{ab}$ and $\int n_a n_b n_c n_d \, d\omega = (4\pi/5)\delta_{a(b}\delta_{cd)}$, this finally becomes

$$\dot{E} = -\tfrac{1}{5}\dddot{Q}^*_{ab}\dddot{Q}^{*ab}. \tag{15.10}$$

It is left understood that when necessary the time average of the expression on the right is to be taken.

Now our own principal interest in (15.10) lay in its application to the behavior of a two-body system such as a binary star. It will be amply sufficient to contemplate for this purpose a pair of bodies P and Q which move in the usual Keplerian orbits according to the equations of the tenth lecture. The actual secular perturbation due to the energy loss to be calculated will be so small as not sensibly to affect the latter. To keep matters as simple as possible I shall take the orbits to be circular. Let P and Q have masses M_1 and M_2, respectively, their coordinates being $x_1{}^a$ and $x_2{}^a$. If their mutual center of mass is taken to be at the origin of coordinates and $z^a := x_1{}^a - x_2{}^a$ are the relative coordinates of P and Q, $x_1{}^a = (M/M_1)z^a$ and $x_2{}^a = (M/M_2)z^a$, where $M := M_1 M_2/(M_1 + M_2)$ is the reduced mass of the system. Then

$$Q^{ab} = \sum_\alpha M_\alpha \left(x_\alpha{}^a x_\alpha{}^b - \tfrac{1}{3}\delta^{ab} r_\alpha{}^2 \right) = M\left(z^a z^b - \tfrac{1}{3}\delta^{ab}\delta_{cd} z^c z^d \right).$$

The last term on the right can be omitted since it disappears on forming time derivatives. If R is the fixed distance apart of P and Q, and ω is their period of revolution, $z^1 = R\sin\omega t$, $z^2 = R\cos\omega t$. The evaluation of $\dddot{Q}^{ab}\dddot{Q}_{ab}$ is now elementary: it turns out to have the constant value $32M^2R^4\omega^6$. Thus

$$\dot{E} = -\tfrac{32}{5}M^2R^4\omega^6, \tag{15.11}$$

retardation and time averaging being redundant here.

As a matter of interest we can insert in (15.11) values for the earth's motion around the sun. Then $M \approx$ mass of the earth $\approx 1.5 \times 10^{-11}$ second, $R \approx 5 \times 10^2$ seconds, $\omega \approx 2 \times 10^{-7}$ second^{-1}, so that $\dot{E} \approx -5.8 \times 10^{-52}$. In more homely units—multiply by $10^{-7}c^5/k$—the rate of energy loss is therefore about 20 watts. This is so ludicrously small that we must search elsewhere if (15.11) is to be compared with the results of observation. Plainly one's best hope is to find a close binary both components of which are very dense, or to all intents and purposes a pair of neutron stars, S_1 and S_2, revolving rapidly about each other. If at least one happens to be a pulsar, radioastronomical measurements of so high a degree of precision could be made that one might hope to find the change in the orbital period of the system.

To get the relevant equation we may use the fact that $E = -M_1 M_2 / 2R$ and $\omega^2 R^3 = M_1 + M_2$ in (15.11). Then

$$\dot{\omega} = \tfrac{96}{5} M_1 M_2 (M_1 + M_2)^{-1/3} \omega^{11/3}. \tag{15.12}$$

If for the purpose of a rough order of magnitude calculation we take M_1 and M_2 both to be about one and a half times the solar mass and $\omega \approx 2 \times 10^{-4}$ second—for reasons which will appear in a moment—then the rate of decrease of the binary orbital period T is of the order of $\dot{T} \approx -2 \times 10^{-13}$. Bearing in mind the astonishing refinement with which pulsar observations can be made, this begins to look hopeful; the more so since if the orbital eccentricity is nonzero the right hand member of (15.12) takes an additional factor $(1 + \tfrac{73}{24} e^2 + \tfrac{37}{96} e^4)(1 - e^2)^{-7/2}$ which increases rapidly with e^2.

Yet, let us reflect! Can we really take (15.12) seriously? In the light of several questions already raised in the ninth lecture it looks as if at best it might be taken as a heuristic guide to the order of magnitude of $|\dot{\omega}|$ to be expected for realistic systems, for there are just too many objections to its derivation. Let us just glance again at a few of them. (1) As mentioned earlier, in the linear approximation each of S_1 and S_2 moves as if the other were not there, yet—inconsistently—we took Kepler orbits to be a first approximation to their actual motion. (2) Use of the linear approximation implies the assumption that nearly Euclidean conditions prevail everywhere, yet this is certainly not the case within S_1 and S_2. (3) The question "what is the linear approximation approximating?" was raised during the preceding lecture. To answer it in the present context, one would have to be able to show that the process which yields the linear approximation as a first step can be continued to arbitrarily high order; that it converges; and that, moreover, it converges to a g-T pair which can be interpreted as representing a physical system sufficiently similar to the system of interest. (4) To the extent that a demonstration of convergence is extant one would at least have to show that higher order terms do not contribute to $\dot{\omega}$ amounts of the same order of magnitude as that on the right of (15.12); and so on. Furthermore, as one goes through the details of particular procedures new difficulties continually arise. At any rate, there can be no question that the whole problem is so far from being simple that one should surely abandon the idea of dealing here any further with its theoretical aspects.

What of observation, however? Remarkably, it seems that a system of the kind we are contemplating has in fact been found. It is usually referred to as *PSR*1913 + 16 and it consists of two neutron stars, each with a mass about 1.4 times that of the sun, and one is a pulsar with a period of oscillation of 59 milliseconds. The orbits are highly elliptical with an eccentricity $e \approx 0.62$

and semimajor axis $a \approx 2.9$ seconds. It has been possible to measure a large number of distinct parameters characterizing the system and their values are mutually consistent. Amongst these is a very large precession of periastron of 4.2 degrees per year—compare this with the precession of Mercury's perihelion!—and this agrees very well with the predicted value. Most importantly, the binary orbital period $T = 2.79 \times 10^4$ seconds has been found to decrease at the rate $\dot{T} \approx -2.5 \times 10^{-12}$ which agrees to within a few percent with the value predicted according to (15.12). Evidently, if the theoretical interpretation of the empirical results is correct, S_1 and S_2 in fact *approach* one another.

Now, one swallow does not make a summer. Observations on a simple system can be afflicted with the consequences of special features inherent in it, of bias or misinterpretations, and so on. Here one has to contemplate the possibility of perturbations of the orbits of S_1 and S_2 being caused by the exchange of matter between them, by the presence of diffuse matter, by tidal effects, and the like. Moreover, bearing in mind that the relevant observations consist exclusively of the measurement of time delays, the value of $\dot{T}$ might merely be an artifact of the common motion of S_1 and S_2 about a third distant, massive component S_3. Be that as it may, the degree of similarity between the observed value of $\dot{T}$ and that calculated from (15.12) is surely very remarkable, the more so as independently of this the precession of periastron is also correctly predicted by the simple binary star model. Whether these coincidences are fortuitous only future observations on other systems of a similar kind can tell. Should they turn out to be genuine, then only (15.12), but not the procedures which led to it, will have been validated. In that case one will have to look for a sound derivation of this result, one which abandons any appeal to naive linearization on the one hand and to energy complexes on the other and which, in fact, avoids any redundant reference to "radiation" altogether. Some very recent work, based on the method of matched asymptotic expansions, appears to have made some progress in this direction.

At this point we might rest content and put further formal work aside. Now, expositions of general relativity theory commonly end with some account of the cosmological problem. The intention seems to be to present an "application" of the theory, much as one might illustrate the operation of the principles of quantum mechanics with an extensive discussion of solid state physics or nuclear physics or whatever. One does not normally do this, however: one regards such topics as effectively distinct disciplines in the sense that they involve a great deal of specialized knowledge, both with regard to their empirical content and to the theoretical and conceptual devices peculiar to them. Cosmology, too, is a study of this kind—already the mere astrophysical knowledge required to do full justice to it is enormous—and to deal with it here seems to me to be out of place. Further, it has special features which reinforce this point of view. First, cosmology deals with the world as a whole, or, more carefully, with the study of the distribution of matter on the largest scale. It therefore has to manage in *fact* without one of the mainstays of physics, namely, the repeatability of observation and experiment. All our observations of necessity are made once, here and now. We cannot alter the world and observe it again. Second, cosmology does not have to be circumscribed by Einstein's theory. In the first instance one might proceed merely taxonomically, so to speak, by making a catalog of all the results of observation; observation, by the way, which has hitherto consisted almost exclusively of the registration of electromagnetic radiations. To interpret them in terms of distant sources one then has to have some kind of theory about the propagation of light and one has to make assumptions about spacetime. Second, even if on the grounds of *P*3, coupled with superficial inspection of empirical results, one endows spacetime with a specific metric which is generically of the kind to which we are accustomed, it is not mandatory to go further and require this metric to satisfy the equation of consistency (8.1). Third, imposition of this requirement implies an enormous extrapolation of the assumed validity of the theory from the nonlocal to the global level, from the approximately here and now to everywhere at all times. Fourth, this in turn may lead to one's having to contemplate conditions in the remote past vastly different from those for which our usual physical laws have been established— densities, pressures, and so on, may nominally tend to infinity. It is difficult to accept extravagant extrapolation to circumstances in which our "ordinary" laws of physics cannot be expected to retain even approximate validity, under which even the possibility of maintaining basic conceptions which we normally take for granted becomes problematic. A point has been reached where physics ends and it is better to admit this than to pretend that in this way one might be able to make a serious contribution to the age-old metaphysical problems of creation, of beginnings and ends.

Be that as it may, even ignoring the last, perhaps somewhat polemical, point, the case for not going on now to study cosmology in its own right is secure enough. What we must still do, however, is to return briefly to the question of boundary conditions. In the first place, we already noted earlier that without the warrant of observational evidence one cannot say anything about the form the metric might take at points sufficiently remote from a given system. This is bound up with a particular defect of any theory whose main formal ingredient is a set of differential equations, namely, the need to impose ancillary conditions which select from the set of all possible solutions of the equations those which are relevant to a given physical state of affairs. Here, in particular, the equations $R_{ij} = 0$ by no means require the metric to be necessarily Euclidean. For example,

$$ds^2 = -dX^2 - dY^2 + dU\,dV + \left[(X^2 - Y^2)\cos 2U - 2XY\sin 2U\right]dU^2,$$

which is reminiscent of (13.26), turns out to be a genuine vacuum metric: every geodesic is of infinite length in both directions—that of null geodesics is to be measured by an affine parameter—g_{ij} has the correct signature, has nonvanishing determinant, and has continuous second derivatives. This metric and the metric $g_{ij} = \eta_{ij}$ therefore provide two alternative g-T pairs, albeit with $T^{ij} = 0$. One might hold that both are to be rejected since they accommodate no matter of any kind and are therefore physically empty. Still, one will probably concede that there are presumably metrics which closely resemble them but which do accommodate matter of some kind; and the temporal behavior of such matter will be different in the two cases. Only if we are careless could we gain the impression that we encountered no problem of this kind when we first investigated planetary motion. In fact, we imposed ancillary conditions upon the metric—in particular, it was to be static. Likewise, the linearized solution (11.3) of equation (8.7) only comes about when radiative contributions to g_{ij}, that is, those arising from solutions of the homogeneous equation $\Box \hat{l}_{ij} = 0$ are declared to be absent by fiat; and this amounts to the imposition of boundary conditions. To all intents and purposes the metric was *required* to become asymptotically Euclidean. Such a demand requires legitimization and this can only rest on empirical grounds. Of course, the fact that it led to correct prediction itself might be thought to provide it. However, because the material system was "small" we can only conclude that at some sufficiently large, but not too large, value of r, r_1, say, the deviation of the metric from Euclidicity is no longer significant, granted that the motion of test particles is confined to $r < r_1$. In effect the boundary conditions have been replaced by junction conditions at $r = r_1$. However that may be, we may wish to consider "very large" systems and then the empirical validation of assumed boundary conditions will have to be direct. We have no choice but to inquire into the character of the spacetime metric on a cosmical scale.

We shall maintain the spirit of our intention to leave the study of cosmology aside by circumscribing our work with extreme severity in every respect; but at least we shall see how one might arrive at any feasible metric at all. Straight away we take the large scale validity of general relativity

theory for granted. Then, however, a casual glance at our surroundings suffices to show that to obtain a metric which will describe them in detail is an impossible task. The best we can hope to do is to proceed on a crude phenomenological level, to treat the distribution of stars, galaxies, interstellar gas, and radiation as if it were just a continuous medium, "the cosmic fluid." Granted the averaging that this picture implies, there is observational evidence that the distribution of matter, as seen from the earth, is isotropic. This isotropy is shared by cosmic radiations incident upon the earth. Certainly it is striking that this should be the case, so much so that—partly in pursuit of *P3*—one adopts a stance of super-Copernican humility as follows. The notion that we just happen to be in a particular, privileged position such that with respect to it isotropy obtains is declared to be unacceptable. Instead one goes to the opposite extreme with the proposition that there are *no* privileged positions. More precisely, a survey of the universe, carried out by any observer at rest relative to the cosmic fluid in its immediate neighborhood, will find isotropy to obtain. The validity of this so-called cosmological principle is obviously totally beyond the possibility of direct verification, but it has testable implications. One certainly has to be careful with the adoption of "principles" which come dangerously close to being metaphysical in character. In the present context one would otherwise be tempted to go one step further by supplementing the cosmological principle with the proposition that no matter *when* the surroundings are surveyed, they will always present the same aspect. This strengthened principle has indeed been put forward in the past, but, conflict with observation apart, it is incompatible with general relativity theory, unless one is willing to admit terms in T^{ij} which represent the local spontaneous creation of matter.

To find a form of the metric in harmony with the cosmological principle we can circumvent the use of mathematical theorems by pursuing an argument possessing a more physical character, as follows. Let O be any selected event. Then, bearing isotropy in mind, choose a coordinate system $(x^1, x^2, x^3, x^4) \equiv (r, \theta, \phi, t)$ with origin at O, such that the metric takes the form

$$ds^2 = -e^{\alpha} dr^2 - r^2 e^{\mu} d\omega^2 + 2f \, dr \, dt + e^{\nu} dt^2, \qquad (15.13)$$

where α, μ, f, and ν are functions of r and t. The motion of the cosmic fluid K is radial, $u^i = (u^1, 0, 0, u^4)$, so that by a transformation involving r and t alone, u^1 can be reduced to zero. One metaphorically calls the new coordinate system co-moving. The generic character of the metric is still given by (15.13). A transformation of the kind $r = r_1, t = \gamma(r_1, t_1)$ can be made to diagonalize the metric without disturbing the co-moving character of the coordinate system. This means that in effect one can simply set $f=0$ in (15.13).

Now, if a test particle is initially at rest with respect to K at some point A it must remain so, since a direction would otherwise be singled out at A, in conflict with the assumption of isotropy. The permanent constancy of its coordinates x^a reduces the geodesic equations to $\Gamma^a{}_{44}(dt/ds)^2=0$, or $\nu'=0$. e^ν is thus a function to t alone, so that it can be removed from (15.13) by the substitution $t_1=\int e^{\nu/2}\,dt$. In effect one can simply set $\nu=0$ in (15.13). Note that one now has $ds=dt$ for any clock at rest in K. Next, let two neighboring elements of K have fixed coordinate differences $(\delta r,\delta\theta,0)$. The fractional rate of increase of their mutual distance is $\frac{1}{2}\partial[\ln(e^\alpha\,\delta r^2+r^2e^\mu\,\delta\theta^2)]/\partial t$. Isotropy requires this to be independent of $\delta r,\delta\theta$ and this will be the case if and only if $\dot\alpha=\dot\mu$, that is, $e^\alpha=e^\mu\zeta^2$, where ζ is a function of r alone, dots denoting derivatives with respect to t. The transformation $r_1=\exp\int\zeta(r)\,dr/r$ in effect removes this from ds^2. In sum, none of the various transformations have disturbed the co-moving character of the coordinate system and (15.13) has so far been reduced to

$$ds^2=-e^{\mu(r,t)}(dr^2+r^2\,d\omega^2)+dt^2. \tag{15.14}$$

The field equations have not yet been used. The stresses in K can only consist of an isotropic pressure and, since $u^i=(0,0,0,1)$, the energy tensor is $T_i^j=\mathrm{diag}(-p,-p,-p,\rho)$. For (15.14) the equation $T_1{}^4=0$ is simply $\dot\mu'=0$, so that generically $\mu=2\ln R(t)-2\ln S(r)$. The remaining field equations are

$$8\pi T_1{}^1+\lambda=-\tfrac{1}{2}e^{-\mu}\left(\tfrac{1}{2}\mu'^2+2\mu'/r\right)+\left(\ddot\mu+\tfrac{3}{4}\dot\mu^2\right)$$

$$8\pi T_2{}^2+\lambda=-\tfrac{1}{2}e^{-\mu}\left(\mu''+\mu'/r\right)+\left(\ddot\mu+\tfrac{3}{4}\dot\mu^2\right),$$

$$8\pi T_4{}^4+\lambda=-e^{-\mu}\left(\mu''+\tfrac{1}{4}\mu'^2+2\mu'/r\right)+\tfrac{3}{4}\dot\mu^2. \tag{15.15}$$

Since $T_1{}^1=T_2{}^2$, the first two show that $rS''-S'=0$, so that $S=a+br^2$, where a and b are constants of integration. After absorbing a factor a^{-1} in R and suitably changing the scale of r, the metric finally emerges in the form

$$ds^2=-R^2(t)\left(1+\tfrac{1}{4}\eta r^2\right)^{-2}(dr^2+r^2\,d\omega^2)+dt^2, \tag{15.16}$$

where $\eta=1$, 0, or -1, as the case may be. It is known as the Robertson-

Walker metric and it accommodates the cosmological principle. One can also arrive at it without using the equations (8.1), but having assumed them to be valid we can now find expressions for ρ and p from (15.15):

$$8\pi\rho = 3\left(\dot{R}^2 + \eta\right)/R^2 - \lambda,$$

$$8\pi p = -\left(2R\ddot{R} + \dot{R}^2 + \eta\right)/R^2 + \lambda. \tag{15.17}$$

That ρ and p depend on t alone is a reflection of the spatial homogeneity of this spacetime which has not had to be separately assumed. Equation (15.17) are of course consistent with the vanishing of $T^j_{i\,;j}$ which implies here only the one relation

$$\dot{\rho} + 3(\rho + p)\dot{R}/R = 0. \tag{15.18}$$

Empirical validation of (15.16) and (15.17) effectively involves all the resources of observational cosmology and theoretical astrophysics. Having agreed, for good reason, not to enter into these subjects, I content myself with the remark that consistency between theory and observation can be attained. It emerges in particular that $R(t)$ is an increasing function of t which, nominally, vanishes at a certain finite time t_0 in the past, ρ and p becoming infinite as $t \to t_0$. This is the kind of singularity to which the reservations I expressed earlier in this lecture referred. Still, one cannot easily escape the conclusion that the universe was much denser and hotter in the past than it is now. Nor can one sustain the claim that the appearance of singularities is merely an artifact of the *exact* symmetries implicit in the assumed distribution of the cosmic fluid, the actual universe being obviously irregular in its details; for there are the so-called singularity theorems which prove that there will necessarily be spacetime singularities also under much more general conditions. At any rate, for our purpose we accept (15.16) as the metric of the smoothed-out universe.

LECTURE 16

Before continuing with general issues, it seems pertinent to digress again upon the notions of distance and of energy, this time within the specific context of the Robertson-Walker metric, the actual form of the function $R(t)$ being supposed known. To begin with, consider the wavelength of light emitted by an atom in some galaxy as it is observed on the earth. Picturesquely speaking, let the appearance of consecutive wavecrests be the two events (r_g, t_g) and $(r_g, t_g + \delta t_g)$, fixed angular coordinates being suppressed. Their reception on the earth constitutes two corresponding events $(0, t_e)$ and $(0, t_e + \delta t_e)$. For a radial null geodesic, from (15.16),

$$\int_0^{r_g} dr / \left(1 + \tfrac{1}{4}\eta r^2\right) = \int_{t_g}^{t_e} dt / R(t) = \int_{t_g + \delta t_g}^{t_e + \delta t_e} dt / R(t). \qquad (16.1)$$

By inspection, $\delta t_e / R(t_e) = \delta t_g / R(t_g)$ so that the shift in wavelength $z := (\lambda_{ge} - \lambda_{gg})/\lambda_{gg}$ is given by

$$1 + z = R(t_e) / R(t_g), \qquad (16.2)$$

t_e being related to t_g and r_g by the first member of (16.1).

One observes in fact that, roughly speaking, the light from fainter and fainter galaxies exhibits greater and greater redshifts. To conclude qualitatively that $\dot{R}(t_e) > 0$ one has to assume that increasing faintness reflects increasing values of r_g. To be able to formulate quantitative relations, more than this is needed. One assumes that any galaxy or other source which is recognizably of a certain definite generic type has a corresponding definite, known absolute luminosity $L(t_g)$. The power $l(t_e)$ incident normally per unit area at the earth—l is the apparent luminosity—may then be calculated. If $D_l := (L/4\pi l)^{1/2}$ it turns out that

$$D_l = r_g \left(1 + \tfrac{1}{4}\eta r_g^2\right)^{-1} R^2(t_e) / R(t_g). \qquad (16.3)$$

146

Since L is supposed to be known and l can be measured directly, D_l is an observable quantity, *called* the "luminosity distance." This terminology reflects the traditional astronomical practice of using the inverse square law of illumination to "measure distance." Bearing in mind that t_g is a known function of t_e and r_g, (16.3) relates the unknown value r_g of r to D_l. By the same token z is a function of r_g alone. Elimination of r_g thus leads to a theoretical luminosity distance-redshift relation which is testable by direct observation.

What is to be noticed here is the arbitrary preference accorded to D_l. Traditional practice is equally well reflected in the so-called distance by apparent size D_a, which is the ratio of the diameter of the source—like its absolute luminosity this is assumed known—to its observed angular diameter. It turns out that

$$D_a = D_l \big[R(t_g)/R(t_e) \big]^2, \qquad (16.4)$$

and it gives rise to another distance-redshift relation. Yet other kinds of distance are often talked about, such as the proper distance

$$D^* = R(t_e) \int_0^{r_s} dr / \big(1 + \tfrac{1}{4}\eta r^2 \big). \qquad (16.5)$$

This, however, is an example of a quantity which is operationally fanciful, to say the least. To determine it, a chain of observers would have to be located radially, each sufficiently close to his neighbor. By prior agreement each would have to measure his usual "infinitesimal" proper distance δl from his nearest neighbor just when the universe presents that aspect to him which it presented to the earth-observer when $t = t_e$. The sum of all these distances would then be D^*.

The various "distance" functions D which have been proposed all reduce to $D = R(t_e) r_g$ when r_g is sufficiently small. Then also $z = H r_g$, where $H := \dot{R}(t_e)/R(t_e)$ is the Hubble parameter. Therefore one has the linear distance-redshift relation $z = HD$. If one *defines* a speed of recession $v :=\partial D/\partial t_e$, this becomes $v = HD$. All this, however, is scarcely relevant to distant sources. Then only D_l and D_a remain viable alternatives, but why should one think of one as "the distance" rather than the other? Why, indeed, should one define the speed of recession simply as the t_e-derivative of one kind of distance or another? There is no unqualified answer: D_l and D_a both accord in some manner with an intuitional notion of distance, but they yield distinct and usually complicated "distance"-redshift relations; the empirical determination of each is practicable and, in principle, straightforward; and each fits easily into the theory. However, in saying that they fit in with our intuitive notion of distance, one must not overlook the fact that in one way or another they may run counter to them. Thus one might be misled by habit into taking it for granted that distances between sources lying on a common light path are additive; but there is no "natural" way to ensure the additivity of D_l.

All this only spells out more specifically what I previously said in more general terms. Whatever distance functions one chooses to define, no unconditional preference can be given to any one of them. It might therefore perhaps be best to avoid talking about distances altogether. In the cosmical context, at any rate, they serve no purpose, apart, perhaps, from one of mere verbal convenience, though even that is doubtful. The theoretical z-D_l relation effectively appeared without reference to distances of any sort; and this is a good example of their redundancy. Moreover, having defined a speed of recession $\partial D/\partial t_e$ one might be tempted to go on and define a corresponding acceleration $\alpha := \partial^2 D/\partial t_e^2$. However, all such quantities have at best some kind of kinematical significance. A test particle is by definition free and its acceleration, in its proper, original sense is always zero. If it is co-moving, it is so permanently, but α need not vanish. A nonvanishing α is, however, not to be ascribed to the action of a force. There is no contradiction: one has gone over from a properly defined local quantity to some other nonlocal quantity to which the same name has been improperly attached.

The second part of this digression deals with the notion of energy in the specific context of a Robertson-Walker universe which has $\eta = 1$, filled with electromagnetic radiation, or any substance whose equation of state is nearly enough $\rho = 3p$. In this case, (15.18) implies the constancy of ρR^4. The total substantial energy m is its density integrated over the elements δV of proper volume. This integral converges, so that m is ρR^3 to within a numerical factor. Therefore $mR =$ constant, which shows that the total substantial energy is not conserved. On the other hand (15.18) may be written as $(\rho\delta V)^{\cdot} + p(\delta V)^{\cdot} = 0$, which is the equation governing the local conservation of energy. Thus energy is conserved locally but not globally. One will regard this as a contradiction only if one draws a careless parallel with an ordinary thermodynamic system. There energy lost in consequence of work done by the pressure reappears in the surroundings, but here the idea of surroundings makes no sense. By the same token, even if the meaning of m is rather formal, no active counterpart M to it can be defined at all. What possible physical meaning could one attach to an "energy momentum P_i of the universe"?

The equation of local energy conservation came from the identity $T_i^j{}_{;j} = 0$, yet it was just this which was previously the starting point for the introduction of an energy complex. One concludes that under the present circumstances no such quantity should be contemplated. This is consistent with the fact that conditions of the kind (14.7) cannot now be formulated; and if one brutally attempts to calculate with $t_i{}^j$ one will not obtain meaningful results. These considerations further vindicate the cautious attitude which we adopted from the start toward the notion of overall energy and its conservation.

It is time to return to one of the motivations for considering the cosmical metric at all, namely, the need for ancillary conditions which will select from among all possible solutions of (8.1) those which we consider to be relevant. What is to be understood by "relevant" here is not quite clear, but let that go for the moment. The one concrete conclusion at which we have arrived is that boundary conditions which involve asymptotic Euclidicity— "Euclidicity at infinity"—must, strictly speaking, be rejected, along with any inferences based on them. On the other hand we must bear in mind that the cosmical metric functions only as a background, for exact spatial isotropy leaves no room for the description of "ordinary" systems or radiative processes at all. These must be represented by perturbations of the cosmical metric and the question of boundary conditions reappears in another form.

A point to which I want to draw particular attention is this: I hold the view that questions concerning the choice of ancillary conditions and of topological alternatives which present themselves in the global interpretation of the metric—I leave these aside here—are to be resolved on empirical, not *a priori*, grounds. In particular, the often repeated contention that Euclidicity at infinity is a "natural" boundary condition is to be rejected. Why is it natural? In the light of empirical evidence it is far from being natural. Concomitantly, a solution of (8.1)—or in preferred terms, a *g-T* pair—is "relevant" if it is of a kind which recognizably describes the world and its parts as we in fact know them. Taking $\lambda=0$ for a moment, any concern at the existence of alternative *g-T* pairs, both with $T_{ij}=0$—we met an example last time—is misplaced: neither is acceptable.

There is, *in fact*, only one metric, that is, one spacetime. To discover it we have no option but to proceed in the language of temporal evolution, for, metaphorically speaking, each one of us must reason outward from his present to the past and to the future. We use field equations to this end—that is their raison d'être—but because of their local character they cannot achieve their purpose without global supplementation. That this should be necessary has sometimes been held to be a severe epistemological defect of the theory, apparently on the grounds that it is somehow not in harmony with the spirit of relativity theory in as far as an absolute element is being introduced into it. It is not easy to understand this objection; the globally absolute is merely a reflection of the uniqueness of the world as a whole and there is no conflict with the requirement that local absolute[II] elements be absent from the theory. What really seems to be at stake here is that the need for ancillary conditions appears to run counter to the notion that "the metrical energy tensor T_{ij} should by itself fully determine the metric g_{ij}."

We are reminded of the Machian ideas to which I referred earlier without comment. These were put forward prior to the development of relativity theory. Partly for this reason they have given rise to much interpretation and reinterpretation. Within the present context I can therefore do no more than to cast a very brief backward glance at them, ignoring all historical perspectives. Very briefly indeed, Mach's position contains three elements, as follows. $M1$: he denies that one can make meaningful assertions about the behavior of matter in a world radically different from the actual world; $M2$: he draws attention to the fact that the plane of oscillation of a Foucault pendulum at the north pole maintains a fixed direction relative to the distant stars; $M3$: he denies that the inertia of a body, that is, the inertial force which balances an external force acting on it, is evidence of acceleration relative to some substantival space, but ascribes it to acceleration relative to distant matter. All this is frequently telescoped into what is loosely called Mach's principle, but since Mach never stated it there are many versions of it, some of them adapted to later theories. To the extent that one does not diverge too far from the original ideas, one encounters statements of the following kind. Ma: inertial frames are not those fixed in some absolute[II] substantival space but are determined by distant matter in the universe; Mb: the inertia of a body is determined by distant matter in the universe. An early example of a subsequent postrelativistic adaptation we have already encountered, namely Mc: T_{ij} by itself alone completely determines g_{kl}.

As a paramount example of the intended operation of $M1$, consider a system consisting of two bodies rotating about each other at the end of a string which we know to be under tension. $M1$ denies that in a world devoid of all other matter one could sensibly maintain that there is any tension in the string on the grounds that states of rotation or nonrotation are now indistinguishable. $M2$ is merely a statement of an observational fact. However, whereas Newton would have regarded it as evidence that the system of distant stars provides an inertial frame by reason of its nonrotation relative to substantival space, Mach takes the view that the distant stars *must* provide an inertial frame since—as asserted by $M3$—it is acceleration relative to these stars which causally leads to the appearance of inertial forces. The nature of these forces is not explained. All this is summed up fairly well by Ma and less explicitly by Mb; but the language used is of course nonrelativistic throughout. There is talk of substantival space, not spacetime; the inertial frames are global; and so on. Moreover, we have good reason to believe that $M2$ is defective in as far as the two periods of rotation are in fact unequal because of the frame-dragging which we encountered in connection with the Kerr metric. On the other hand, if $M2$, and Ma with it, is merely taken as leading to the suggestion that inertial frames are not given absolutely, but are somehow causally related to the distribution of matter, then the general theory of relativity accommodates it. Yet objections are frequently raised which declare that this accommodation is only partial, in the sense that, contrary to Mc, a given distribution of matter is consistent with distinct distributions of local inertial frames. This is true if once concedes the admissibility of counterfactual boundary conditions—recall $M1$ here, by contrast—or if one holds that no boundary conditions should be required at all. On the other hand, such positions will be rejected if one strictly takes scientific enterprise to consist of the discovery of the behavior of the actual world. The formulation of general laws of physics is part of this enterprise, but it may be incomplete—for example when laws are presented as differential equations without boundary conditions. They will then appear to be capable of governing possible alternative worlds. Whether one should simply reject all possibility talk here as meaningless is arguable; one would certainly have to do so if one insisted on empirical verifiability as a criterion of meaningfulness.

Beyond this I shall confine myself to a few remarks about Mb. It has been taken to imply that if one were to remove all matter from the universe other than one or two particles then their inertia would be, nearly enough, zero. Such a claim seems to me to border on the absurd. Not only is it totally at variance with $M1$, but one must question the meaningfulness of the idea of customary physical laws under such virtually unintelligible circumstances. Still, let that go: the notion survives that the inertia of a body is "caused by" its interaction with distant matter. There have been attempts to construct theories which incorporate this idea; but here I merely point out that general relativity theory does not. The inertia of a body is a measure of its acceleration in response to a force which acts on it. Distant matter and its history, however, merely plays its part—through the equation of consistency (8.1) and boundary conditions which go with it—in determining the local inertial frames to which this acceleration is referred. Nothing survives, however, in the equations of motion, viewed locally, which characterizes the presence of distant matter. In any case, if inertia were to be related to acceleration with respect to distant matter, then what is this acceleration?

Preoccupation with Machian ideas has presumably survived because they marked an important stage in the continuing conflict between substantivalist and antisubstantivalist, or relationist, views. Already earlier in this course I asked some questions in this context, but made no attempt to answer them. Now, too, they will be left unresolved. Let me just remind you that, roughly speaking, the substantivalist claims—in the relativistic context —that spacetime is an entity in its own right, that it has structure, and that it can be said to exist even in the absence of any ordinary material objects at all, matter as usual including here electromagnetic fields and the like. The relationist, on the other hand, denies this, but maintains that spatiotemporal assertions are to be understood as being about spatiotemporal relations which subsist between material objects, rather than the attribution of properties or features to some mysterious entity called spacetime. Of course, one must not take the adjective "substantival" to imply a necessary commit-ment to any substance in the ordinary sense. Where once it might have seemed feasible to think of space as a substance, albeit one which, like Maxwell's aether, had idiosyncratic properties, the idea of spacetime as an "everyday" substance simply does not make sense. "Motion through space-time," the "permanence of spacetime," the "static four-dimensional picture" —these are examples of phrases which merely reveal semantic confusion. Be that as it may, the point I want to make is that general relativity theory does not appear to make a significant contribution toward the resolution of the issue of substantivalism versus relationism; in particular, that spacetime is now not absolute[II] is not germane to it.

We are approaching the end of our course—only one more lecture remains. It will make it easier to deal succinctly with some points then to be reviewed if I talk briefly now about the idea of universal and differential forces and the distinction between them. This has so far not been brought out explicitly and it may help to strengthen our insight. What I am about to do is allegorical and any artificiality may safely be ignored.

Let D be a given plane-sided disk consisting of some material of which it is known that its thermal coefficient of expansion is zero. Remember that conditions are here supposed to be entirely classical and the thermal rigidity could be verified by means, say, of stretched strings passing between various points on D. In like manner it will be possible to draw a Cartesian coordinate grid x, y on D. Now imagine a physicist M to be instructed to determine the spatial relationships between points on D on the condition that the only means available to him for this purpose are metallic measuring rods. Furthermore, he is supposed to be entirely ignorant of thermal phenomena.

Now, with our superior knowledge we can say what he will find, taking the thermal condition of D into account, the temperature distribution being supposed stationary. Suppose that any given rod has a constant coefficient of expansion β, so that, if l_0 is its length at the origin O, it is $l = l_0(1 + \beta\tau)$ at the point $P[x, y]$, where τ is the difference in temperature between P and O. Then M will assign to the points (x, y) and $(x+dx, y+dy)$ on D the mutual distance ds given by

$$ds^2 = \psi(dx^2 + dy^2),\tag{16.6}$$

where $\psi := [1 + \beta\tau(x, y)]^{-2}$. He, of course, knows nothing about τ, but he can empirically determine ψ. He can further calculate the two-dimensional Riemann tensor, which effectively reduces to the single component R_{1212} and will generally find it to be nonzero: he is stuck with a non-Euclidean metric. But is he? No, for he may argue that according to his preconceptions, at any rate, the disk *must* be Euclidean. He infers that there is some mysterious "force" at work the nature of which he doesn't understand, but which causes an increase in length of the rod; in other words, a rod at the origin, nominally of unit length there, is expanded by the force so that its length will be $1/\sqrt{\psi}(x, y) =: \phi(x, y)$ after transport to the point (x, y). In fact, M might go so far as to insist that the force is just the fractional change in length to within a universal constant factor, $f := \alpha(\phi - 1)$, since he feels that to bring elastic properties into the picture would be an unnecessary complication. Of course, being a competent scientist, M will not be content to use only rods made of a particular metal, and he will repeat his investigations using rods made of some other metal. He finds this time that in place of ψ in (16.6) he has a different function ψ_1, say, so that this time the force is $f_1 = \alpha(\phi_1 - 1)$. Upon investigating the results obtained he finds that $f/f_1 = \beta/\beta_1$, which makes him feel that he is on the right track. He can ascribe to every rod a "charge" β, the force then being the product of this charge with a function characterizing D alone. As a final check he takes two rods made of different materials, of equal length when at O, and transports them to P, where he compares their lengths directly. In general he finds these lengths to be distinct, the difference being moreover independent of the rate of transport. In short, f is a differential force in as far as it distinguishes between different materials.

What, however, if for some reason all metals had the same coefficient of expansion? ψ would then be, for D, a universal function and f a universal force in that all rods would be influenced in the same way. In particular, any two rods, if equal in length at O would still be congruent at any arbitrarily chosen point P. What then is the point of maintaining that their lengths change under transport? By no direct comparison can that assertion be verified by M. Furthermore, a force has had to be introduced for the sole purpose of "explaining" this unverifiable behavior. Here unverifiability is to be taken in the sense that verification cannot amount to more than showing that it leads to the rescue of the Euclidean form of ds^2; but why should one wish to rescue it, seeing that one merely satisfies an arbitrarily imposed *a priori* demand? Being sensible, *M* will rest content with (16.6) as it stands, on the grounds that the introduction of a universal force is thus avoided. In this way he adheres to the principle of conceptual parsimony—a concomitant of *P3*—which surely plays an important part in determining the acceptability of any scientific theory.

We could go on to construct an allegorical counterpart to equation (8.1), but it is not worth our while. The important point at issue is that when a theory contains universal effects these should be eliminated by a transformation of the theory. If, for example, one maintains the overall Euclidicity of spacetime one has to have a theory in which free particles are in general accelerated relative to a frame fixed with respect to the distant stars, but the acceleration is independent of their nature—picturesquely speaking it does not matter whether they are made of lead or of silver. The general theory of relativity eliminates this universal effect among others. Even granted that there is no absolute compulsion to do this, a Euclidean theory designed to describe the motion of matter correctly would be extraordinarily complex. Nevertheless, because, as we have already learned, any answer to the question whether spacetime is Euclidean or not is subject to prior agreement on various conventions, it has sometimes been asserted that the first choice is to be maintained at all costs. This curious stance seems to be connected with the notion that non-Euclidean spatial structures cannot be visualized. Is this acceptable? Not only are there good reasons for believing "spatial intuition" to be topological rather than metrical in structure, but in any case, why should the ability to visualize—a matter of psychology—be regarded as basically relevant to physics?

LECTURE 17

Let me begin this final lecture with a remark on a feature of the whole course of which you will no doubt have become aware long ago. It is this: hitherto I have studiously avoided the use of two terms, namely, gravitation and curvature, which are normally ubiquitous in accounts of general relativity theory. My purpose in doing so was entirely didactic, not the capricious achievement of a mere tour de force. In short, it seemed desirable to avoid linguistic confusion from the outset. Reflect that on the theoretical level one has to contend with several distinct languages or vocabularies, L, each with its conflicting counterpart L^*. Strictly speaking they are merely alternatives which go toward characterising possible languages in which the theory as a whole may be formulated. Their enumeration naturally takes place against the background of the material of preceding lectures.

$L1$ rejects the presence of universal forces. In the first instance free particles and test light pulses provide the kinematical basis to which the dynamical motion of other matter is referred. All spatiotemporal relations are subsumed under a physical nonabsolute geometry which is a realization of an abstract non-Euclidean geometry. Test objects are subsequently recognized as redundant. They may therefore be abandoned, whereupon the vagueness associated with local inertial frames becomes irrelevant. L^*1 is adapted to the view which arbitrarily assigns to spacetime a Euclidean structure. To account for the fact that independently of the choice of reference system the tracks of free particles and light pulses in general do not realize the straight lines of Euclidean geometry, universal forces have to be admitted, so-called gravitational forces. From the standpoint to which $L1$ is adapted these are an illusion, within $L1$ itself they do not occur. $L2$ reflects the belief that emphasis on the terminology of pure geometry is neither necessary nor desirable, while L^*2 rejects this. $L3$ is appropriate to a descriptive enterprise which takes spacetime as a datum and histories as wholes, whereas L^*3 is pertinent to the idea of temporal evolution, spacetime in some way being split up into two subspaces, one spacelike and one timelike. The enumeration could be extended. For instance, one might distinguish between $L4$ which takes the substantiality of spacetime for granted and $L*4$ which does not; and so on. At any rate, in these lectures I have attempted to adhere strictly to $L1$ and $L2$, eschewing the use of $L*1$ and $L*2$, while making the occasional transition from $L3$ to $L*3$ or vice versa explicit.

Of the various distinctions which have been drawn, that between $L1$ and $L*1$ is, without question, essential. One of them has to be adopted to the exclusion of the other if we are not to be submerged in confusion. We have already seen in detail that $L1$ is to be preferred and it is fair to say that on this point there is almost universal agreement. The abandonment of $L*1$ in favor of $L1$ is sometimes paraphrased as the "geometrization of gravitation." This at least is less strange than the proposition that "physics has become geometry." If geometry here is intended to mean abstract geometry it is false, if physical geometry it is effectively tautologous, for at best it then expresses an arbitrary preference for $L*2$ over $L2$. Be that as it may, there is a rather widespread tendency to regress to the use of $L*1$, perhaps because "gravitational field" is a convenient circumlocution for the field g_{ij}, where "field" now merely stands for something like a set of functions defined on a manifold, without any physical implications that it subjunctively defines forces. In parenthesis, it is often said that there is a "real" gravitational field in a region if and only if the Riemann tensor does not vanish there. Of course one is entitled to talk like this—by way of definition, at any rate—but one is then merely emphasizing that only when the Riemann tensor vanishes will free particles nonlocally pursue Euclidean motions relative to each other. Evidently one is thinking in terms of $L*1$ again.

All this might be regarded as a mere tilting at windmills, were it not for the fact that the simultaneous use of $L1$ and $L*1$ leads to genuine error. In one well-known exposition of the theory one reads that "the g-field characterises not only the gravitational field but the behaviour of measuring rods and clocks as well"; in a second, that "the metric field determines the rate of ticking of atomic clocks"; and so on. Now, within $L1$ it is correct to say that in such and such a region spacetime is not Euclidean and to exhibit a specific metric associated with it. With $L*1$ it is correct to say that the length of a measuring rod or the period of a clock placed in the region vary with their position; but one cannot have it both ways. To return to the allegory of the disk D with its hidden thermal inhomogeneity, that a particular non-Euclidean physical geometry obtains is contingent upon the maintenance of a given measuring rod as a standard of length or distance all over D. In other words, the form of the metric coefficients g_{ij} is determined empirically, given some coordinatization of D and the invariability of the measuring rods which obtains *by prescription*. Once this has been done, one can deduce everything that can be said about the conditions on D. If, however, one lays down the form of the g_{ij} from the outset, say $g_{ij} = \delta_{ij}$, the length of any rod will be found to vary from point to point as it is transported over D. These alternatives exactly represent the use of $L1$ and L^*1, respectively. It would obviously be wrong to speak within $L1$ of "a change in length of a rod which is determined by the g_{ij}." To obtain a definite empirical metric at all, the behavior of measuring rods must first be specified by fiat. One does not necessarily have to prescribe that the rods be invariable, but any alternative will specify their lengths as functions of the coordinates, not of the metric which is as yet unknown, and the form of which depends on the particular specification actually provided.

In the general theory of relativity the circumstances are analogous. In particular, underneath any statement to the effect that *the* empirical metric of spacetime happens to be such and such lies the prescription that clocks of a certain kind are to be taken as invariable. To speak then of varying rates of ticking of clocks or, worse, of a "slowing down or speeding up of time" is an unfortunate obfuscation which should be dispensed with. They are forms of speech sometimes used in discussions of redshifts, for example, but they only hinder their clarification. Verification of the solar redshift should be regarded as evidence that, metaphorically speaking, a cesium clock on the sun does in fact "tick at the same rate" as a cesium clock on the earth; more precisely, that by prescribing the equality of these rates one arrives at a valid description of observed phenomena, in particular of the histories of finitely separated clocks.

Directing our attention now to $L2$ and L^*2, it might be thought to be a matter of indifference whether one gives preference to the one or to the other. Such a view, however, would ignore a number of unattractive features of L^*2. It overemphasizes analogies with structural aspects of everyday things, suggesting that one is thus able to visualize what is in fact not visualizable; it hints that such analogies are essential to an understanding of the theory when the contrary may well be the case—is not the Bohr model a hindrance to an understanding of quantum mechanics?; it gives rise, on the colloquial level, to pretty figures of speech, sometimes misleading, sometimes vaguely self-contradictory; and it pulls the ontological wool over our eyes by trying to habituate us to the unquestioned acceptance of space as a substance.

Rather than subject these various points to detailed analysis, I shall merely say a few words about the typical term "curvature" and its concomitants. Its visual association is with objects encountered daily: in simple terms, one thinks of their surfaces as locally characterized by a measure of the degree to which they visibly diverge from planeness at any given point on them. It is just such visual associations and their implications which can be deceptive. Recall that there are particular branches of mathematics, that is, certain specific sets of axioms and theorems inferred from them, which are called geometries by common consent. The axioms provide relations between primitive terms and it is not unusual to refer to a given abstract system as a geometry only if it admits one of a specified group of transformations under which these relations are invariant. The group then characterizes the geometry. Be that as it may, it is not at all necessary that any physical interpretation of a given geometry must take some standard form. Therefore, if "curvature" is part of geometry $\mathcal{G}$, then it merely characterizes certain aspects of the abstract relations which constitute $\mathcal{G}$, and a particular physical realization may interpret curvature in a way far removed from customary visual associations. Likewise "space" will stand for a specific aggregate of primitive objects which enter into axioms of $\mathcal{G}$. When the curvature measure K is nonzero one may then simply say that the space is curved; but this is no more than a convenient figure of speech. It would perhaps be better to speak of $\mathcal{G}$ itself as being "flat" or "curved" according as K does or does not vanish.

A brief return to the allegorical disk D suggests itself. M, having studied differential geometry, naturally associates with the metric (16.6) the usual Gaussian curvature. This is here

$$K = \phi \nabla^2 \phi - |\text{grad } \phi|^2 \tag{17.1}$$

and he so comes to say that D is curved; granted of course that within the allegory he cannot go outside D in any sense at all. More specifically, suppose M to have found by measurement that $\phi = 1 + \frac{1}{4}a^2(x^2 + y^2)$, where a^2 is a positive constant. *We* know that this corresponds to β being positive and the temperature increasing outward as the square of the distance from the origin—with our measure of distance. M, of course, is ignorant of all this. He merely finds by calculation that $K = a^2$. In other words he finds D to have a constant positive curvature $\sqrt{K} = a$. He therefore refers to it henceforth as part of a two-dimensional spherical space of radius $1/a$, having decided that methodological principles forbid him to appeal to universal forces. In so doing he is going over from a non-Euclidean plane to a Euclidean model of it.

The relevance of all this to general relativity theory should be obvious and I need not enlarge upon it. It is worth saying, though, that it is not only in popular expositions where one finds talk of gravitation bending space or everything in space, of gravitation distorting time, or "spacetime being curved in the presence of gravitation", and so on, as if this were more than metaphor at best. At any rate, I have sought to avoid such unhappy parlance throughout, together with what it might, but should not, suggest, by simply avoiding the words curvature and gravitation altogether.

This is a reminder that the term "source," too, was used nowhere except in some questions about it which were raised in the eighth lecture. Not having accepted the supersubstantivalist position which seeks to establish that all material things are themselves structured pieces of spacetime, any argument about the legitimacy of referring to T_{ij} as the source of the "gravitational field"—that is, g-field—takes on a semantic character. If the metric is static or at least stationary, asymptotically flat, and singular solutions are disallowed, then it is Euclidean or not according as T_{ij} is or is not zero. In this sense one might speak of T_{ij} as the source of the g-field. In the general case, however, g_{ij} need not be Euclidean though $T_{ij}=0$ everywhere and in this sense the g-field is then source-free. Its nonlinearity is irrelevant, as the case of the Maxwell field demonstrates. There one simply calls the four-current the source simply by reason of terminological convenience: the field is or is not source-free according as $f^{ij}_{\;\;,j}$ is or is not zero. The corresponding criterion in the gravitational case is whether P^{ij} vanishes or not.

To ask the alternative question whether the energy tensor is the "cause" of the g-field complicates the issue, in that it lands us with the notorious difficulties surrounding the idea of causation. To deal with them here is plainly out of the question. I shall therefore simply take it that to assert that the energy tensor causes the g-field—or better, causes it to have the structure it has—means that there exists a relationship between them which entails the predictability of the second, given the first. Such predictability in fact does not obtain. It follows that the causal relationship can at best have a much weaker meaning, in that it will have to involve the boundary conditions.

The case for speaking of source or cause at all is further undermined by the following reflection. One thinks of sources, like that of the Maxwell field, as being in principle characterizable independently of the field, while their motions can be freely prescribed. Then, given the boundary conditions, the corresponding field may be found. The situation in the general theory of relativity is quite different. One may be able to give the generic structure of T_{kl}, but it depends explicitly on the g_{ij} as well; mutual distances between bodies cannot be given unless the metric is known; and, above all, their motions cannot be freely prescribed since they must satisfy the condition $T^{ij}_{\;\;,j}=0$. All of this leads to the view that (8.1) should be regarded as a condition of consistency, which is what we have done all along. As for talk about the field being coupled to itself, or, when $T_{kl}=0$, being its own source: these are metaphorical phrases. The first merely expresses the nonlinearity of the field equations, the second at best reflects an arbitrary interpretation of formal features of some of their solutions. At worst it is entirely unacceptable, namely, when "source" is here naturally identified with "cause"; for then one is faced with the quaint proposition that a field is predictable if it is known.

At this stage we shall do well briefly to analyze—perhaps I should say summarize—what lies behind the designation "general theory of relativity." Whatever the compass of the body of knowledge which it covers may be, it is wise to represent it by one symbol, $\mathfrak{R}$, say, if a Pickwickian phraseology is to be avoided here. As once before in the context of the special theory, three laconic questions present themselves: why theory, why general, why relativity? As regards the first, the answer seems to be that $\mathfrak{R}$ functions both as a metatheory and as a theory. In justification, recall that at the start of the seventh lecture three guidelines, embodied in $P1, P2$, and $P3$, were enumerated toward the establishment of the differential equations governing the metric. We shall have a set of guidelines for the formation of $\mathfrak{R}$ as a whole if, in the light of what we already know, we add to these certain stipulations in the form of a further "principle" $P0$. To keep it short, the term "Riemannian metric" is here to be understood as meaning an invariant metric ds^2 which is a homogeneous quadratic form of signature -2 in the coordinate differentials. $P0$ consists of three parts as follows:

$P0$(i):   the special theory of relativity is valid locally;

$P0$(ii):   the structure of spacetime is characterized nonlocally by a Riemannian metric;

$P0$(iii): the coefficients of the metric are not absolute[II], but obey differential equations which relate them to the distribution of matter and energy.

The metatheoretic content of $\mathfrak{R}$ is represented jointly by $P0$(i) and $P1$, but the rest is about a specific theory, namely, of spacetime structure. To this extent, then, one indeed has a conflation of a theory and a metatheory. Because they stand in a symbiotic relation to each other, the distinction between them is not always made clear. One consequence of this is the fashionable view that $\mathfrak{R}$ "is merely a theory of gravitation"; but this leaves the metatheoretic aspects of $\mathfrak{R}$ largely out of account.

On second thoughts one might in fact prefer the view that the compass of $\mathcal{R}$ should be so restricted that only a metatheory remains. The general theory of relativity would then be concerned only with the consequences of imposing regulative principles upon *prior* theories including some theory of spacetime. This is in precise analogy with the view usually taken of the special theory of relativity.

Next, the question "why general?" The qualification "general" stresses that the counterfactual premiss of the indefinite extendibility of local inertial frames is abandoned. In a neighborhood of any event the condition of locality must be satisfied by all inertial frames both individually and with respect to each other if they are to be defined at all. Nonlocally all material frames enjoy like status. Finally, "relativity" again emphasizes that the description of motions requires material referents. To maintain consistency with the absence of any privileged class among these, field equations must be absolutely form invariant under transitions from one reference system to another.

The verbal similarity between this summation and that given earlier for the "special theory of relativity" is deliberate. It is helpful in disposing of the few questions from among the many asked at the very beginning of the course which yet remain to be answered. Foremost amongst them is this: is the general theory of relativity a generalization of the special theory or is it not? Toward its resolution we recall the tentative criteria formulated earlier and ask ourselves first, whether there is a strong family resemblance between their conceptual frameworks; and second, whether the general theory can be regarded as collapsing into the special under certain circumstances. That there is a strong family resemblance will be readily granted. Further, the special theory pretends that inertial systems are global, or, in other words, that "gravitation" does not exist. This corresponds in the general theory to the special situation in which the requirement of asymptotic Euclidicity of the metric is combined with the rejection of all but test systems. All in all, it seems legitimate to regard the general theory as a generalization of the special, the more so as locally the latter survives intact within the former. One is reminded of nonequilibrium thermodynamics in which the assumption of local thermodynamic equilibrium is an essential ingredient.

Yet—not only is the conclusion just reached sometimes vigorously denied, but it is claimed that the general theory is not a relativity theory at all, that the idea of relativity, as it is supposed to be represented within the special theory, is not generalizable. We naturally suspect that we are faced with semantic differences. Now, one is entitled to take "relativity" to refer to the presence of a certain kind of relativity principle which asserts that members of a specified class of reference frames are mutually equivalent; meaning here exactly that if there is a process described by functions $\phi_A(x^k)$ when referred to one frame, there is a possible process which when referred to an equivalent frame is described by the *same* functions $\phi_A(x'^k)$ of the new coordinates. In the special theory all inertial frames are mutually equivalent. They are related to each other by Lorentz transformations and all field equations must be properly form invariant under these. Observations referred to inertial frames can yield no quantity which absolutely distinguishes any one of them. The trouble with all this is that one is talking about a state of affairs which simply does not obtain in the real world and the theory, taken at face value, is physically vacuous.

It is true, of course, but not surprising that, granted the narrow interpretation of "relativity" in terms of the overall homogeneity of spacetime, the general theory is not a relativity theory at all. The metric no longer being absolute its coefficients g_{ij} must occur among the functions $\phi_A(x^k)$. Given $g_{ij}(x^k)$, there is, however, no transformation which will take this into $g_{ij}(x'^k)$ unless g_{ij} happens to admit at least one Killing vector, but the presence of a Killing vector would inject an absolute element into the metric and is therefore irrelevant. To save anything of narrow relativity on a global level one has to take some rather drastic steps such as that of maintaining that asymptotic Euclidicity is the only admissible boundary condiion. It is difficult to see why one should allow oneself to do this without any reference to empirical knowledge.

The conclusion that the relativity theory in the narrow sense is counterfactual provides compelling motivation for the construction of a factual— more "general"—theory which takes cognizance of the *fact* that in a neighborhood of any arbitrarily selected event P there exists an infinity of local, mutually indistinguishable inertial frames, such neighborhoods, corresponding to different choices of P, being likewise indistinguishable by means of local observations alone. In the special theory, with the metric absolute, equivalence of frames can in effect be transcribed into proper form invariance of the laws describing physical processes under transitions from one global inertial frame to another. The general notion of form invariance is still appropriate when the metric is not absolute and the g_{ij} become field functions. Without an absolute metric no nonlocal coordinatization of events can be given preference and concomitantly no nonlocal system of reference enjoys a privileged status. In this sense, all such systems are now equivalent; but it is the absolute character of form invariance which is significant. On the local level the use of an arbitrary reference system in general leads to a description of physical processes which involves the presence of nominal universal forces: nominal, because they can be eliminated—that this is always possible is just the content of the principle of equivalence. They are in fact eliminated by going over to a local inertial frame. The crucial point in all this may be put as follows: a theory is to be regarded as a "theory of relativity" if it is designed to accommodate the proposition that the relation "the body A is at rest relative to B" is a two-term relation the second term, B, of which cannot be "space."

To conclude the course it is fitting that I should provide a list of some readily accessible books which I have found useful in one way or another during the preparation of these notes. It appears here in the form of the bibliography on pages 165–166.

BIBLIOGRAPHY

R. Adler, M. Bazin, and M. Schiffer. *Introduction to General Relativity*, 2nd ed. (New York: McGraw-Hill, 1975).

P. Aichenstein. *Concepts of Science* (Baltimore: Johns Hopkins Press, 1968).

J. L. Anderson. *Principles of Relativity Physics* (New York: Academic Press, 1967).

P. G. Bergmann. *Introduction to the Theory of Relativity* (Englewood-Cliffs, Prentice-Hall, 1942).

M. Berry. *Principles of Cosmology and Gravitation* (Cambridge: University Press, 1976).

H. Bondi. *Cosmology* (Cambridge: University Press, 1968).

B. A. Brody (Ed.). *Readings in the Philosophy of Science* (Englewood Cliffs: Prentice-Hall, 1970).

R. Carnap. *Philosophical Foundations of Physics* (New York: Basic Books, 1966).

C. Clarke. *Elementary General Relativity* (London: Edward Arnold, 1979).

A. S. Eddington. *The Mathematical Theory of Relativity* (Cambridge: University Press, 1923).

A. Einstein et al. *The Principle of Relativity* (London: Methuen, 1923).

H. Feigl and M. Brodbeck (Eds.). *Readings in the Philosophy of Science* (New York: Appleton-Century-Crofts, 1953).

V. Fock. *The Theory of Space, Time and Gravitation*, 2nd ed. (Oxford: Pergamon, 1964).

B. C. V. Fraassen. *An Introduction to the Philosophy of Time and Space* (New York: Random House, 1970).

R. M. Gale. *The Language of Time* (London: Routledge-Kegan Paul, 1968).

A. Grünbaum. *Philosophical Problems of Space and Time* (London: Routledge-Kegan Paul, 1964).

J. C. Graves. *The Conceptual Foundations of Contemporary Relativity Theory* (Cambridge, Mass.: MIT, 1971).

I. Hinkfuss. *The Existence of Space and Time* (Oxford: Clarendon Press, 1975).

W. Israel (Ed.). *Relativity, Astrophysics and Cosmology* (Dordrecht: Reidel, 1973).

L. D. Landau and E. M. Lifshits. *The Classical Theory of Fields*, Vol. 2, revised 2nd ed. (Oxford: Pergamon, 1962).

G. C. McVittie. *General Relativity and Cosmology* (London: Chapman & Hall, 1956).

G. C. McVittie. *Fact and Theory in Cosmology* (London: Eyre & Spottiswoode, 1961).

165

C. M. Misner, K. S. Thorne, and J. A. Wheeler. *Gravitation* (San Francisco: Freeman, 1973).

S. Morgenbesser (Ed.). *Philosophy of Science Today* (New York: Basic Books, 1967).

J. D. North. *The Measure of the Universe* (Oxford: Clarendon Press, 1965).

W. Pauli. *Theory of Relativity* (Oxford: Pergamon, 1958).

A. Papapetrou. *Lectures on General Relativity* (Dordrecht: Reidel, 1974).

W. V. O. Quine. *From a Logical Point of View*, 2nd ed. (New York: Harper & Row, 1961).

H. Reichenbach. *The Philosophy of Space and Time* (New York: Dover, 1958).

W. Rindler. *Essential Relativity*, 2nd ed. (New York: Springer, 1977).

H. P. Robertson and T. W. Noonan. *Relativity and Cosmology* (Philadelphia: Saunders, 1968).

P. A. Schilpp (Ed.). *Albert Einstein: Philosopher Scientist* (Evanston: Library of Living Philosophers, 1949).

M. Schlick. *Philosophical Papers*, Vol. 1 (Dordrecht: Reidel, 1979).

J. A. Schouten. *Ricci Calculus* (Berlin: Springer, 1954).

D. S. Sciama. *The Unity of the Universe* (London: Faber and Faber, 1959).

R. U. Sexl and H. K. Urbantke. *Gravitation und Kosmologie* (Vienna: Wiener Berichte, 1973–1975).

L. Sklar. *Space, Time and Spacetime* (Berkeley: University of California Press, 1974).

P. Suppes (Ed.). *Space, Time and Geometry* (Dordrecht: Reidel, 1973).

J. L. Synge. *Relativity: The Special Theory* (Amsterdam: North Holland, 1964).

J. L. Synge. *Relativity: The General Theory* (Amsterdam: North Holland, 1971).

R. C. Tolman. *Relativity, Thermodynamics and Cosmology* (Oxford: University Press, 1934).

A. Trautmann, F. A. E. Pirani, and H. Bondi. *Lectures on General Relativity* (Englewood Cliffs: Prentice-Hall, 1965).

O. Veblen and J. H. C. Whitehead. *The Foundations of Differential Geometry* (Cambridge: University Press, 1932).

S. Weinberg. *Gravitation and Cosmology* (New York: Wiley, 1972).

L. Witten (Ed.). *Gravitation: An Introduction to Current Research* (New York: Wiley, 1962).

V. D. Zakharov. *Gravitational Waves in Einstein's Theory* (New York: Halsted Press, 1973).

Appendix

ON NOTATION AND
EUCLIDEAN TENSORS

1. Defining relations. An equation $A = B$ either asserts the equality of the quantities A and B previously defined or else serves as a defining relation for either A or B, in which case this is emphasized by writing $A := B$ or $A =: B$, respectively.

2. Indices. The indices labeling coordinates and the components of tensors or other multicomponent objects are printed exclusively in lowercase italic type. Where the range is not explicitly specified it is $1, 2, 3$ when the index is one of $a, b, \ldots, h$ and $1, 2, 3, 4$ when it is one of $i, j, \ldots, z$, with $a, b, \ldots, z$ taken in their natural alphabetic order.

3. Summation convention. When an index occurs twice, once as a subscript and once as a superscript, summation over the appropriate range is implied. They are then "dummy indices." For example, $t^{ia}_{\ jac}$ stands for $\Sigma^3_{a=1} t^{ia}_{\ jac}$, $t^{ia} q_{ibc}$ stands for $\Sigma^4_{i=1} t^{ia} q_{ibc}$. No index can occur more than twice in such an expression. *Contraction* thus amounts formally merely to the identification of a subscript with a superscript.

4. Transvection. By the product of two multicomponent objects one should understand the direct product. For example, the product of s^{ij} and $t^k_{\ lmn}$ is the object the components of which are $s^{ij} t^k_{\ lmn}$. To *transvect* one object with another is first to form their product and then to follow this by one or more contractions. Thus $s^{ij} t^k_{\ imn}$ or $s^{ij} t^k_{\ lij}$ are transvections; and there is no ambiguity in the direction to transvect s^{ij} with $t^k_{\ imn}$.

5. Diagonal matrices. The values of the components of any two-index object can be exhibited as a square matrix. When a matrix M is diagonal and the diagonal elements, taken in order from the top left hand corner are $u, v, w, \ldots$, one writes $M = \mathrm{diag}(u, v, w, \ldots)$. In particular, $\delta_{ab} := \mathrm{diag}(1,1,1)$ and δ_a^b, δ^{ab} likewise. Also $\eta_{ij} := \mathrm{diag}(-1, -1, -1, 1)$ and η^{ij} likewise, while $\delta_i^j := \mathrm{diag}(1,1,1,1)$.

6. Partial derivatives. If f is a function of variables x^i, the partial derivative of f with respect to x^j is denoted by the subscript j following a comma. For example $\partial t^i / \partial x^j =: t^i_{,j}$ or $\partial^2 f / \partial x^i \partial x^j =: f_{,ij}$.

7. Kernel-index notation. If a point P has coordinates x^i and one goes over to a new coordinate system, the coordinates of P become $x^{i'}$, not $x^{\prime i}$. This characterizes the "kernel-index notation." If the x^a-component of some electrostatic field is E_a relative to a coordinate system $\{x^b\}$, then it is $E_{a'}$ relative to a different coordinate system $\{x^{b'}\}$, whereas E'_a would be the x^a-component of a *different* field relative to the *original* coordinate system.

8. Linear transformations. In the Euclidean tensor calculus all coordinate transformations are linear: the typical transformation C is $x^{i'} = L^{i'}_i x^i + L^{i'}$, where the $L^{i'}_i$ and $L^{i'}$ are constants. Evidently $L^{i'}_i = x^{i'}_{,i}$ and this is the definition of $L^{i'}_i$ even under nonlinear coordinate transformation. Note that i' and i are generically distinct indices, so that no summation is implied. The transformations are required to be nonsingular: $L := \det L^{i'}_i \neq 0$.

9. Inverse transformations. The transformations C form a group. If $x^i = L^i_{i'} x^{i'} + L^i$ is the inverse of C, $L^i_{i'}$ is the inverse of $L^{i'}_i$, that is, $L^i_{i'} L^{i'}_j = \delta^i_j$ and $L^{i'}_i L^i_{j'} = \delta^{i'}_{j'}$.

10. Tensors and their valence. By definition, the components of a tensor t of covariant valence p and contravariant valence q, relative to two coordinate systems $\{x^{i'}\}$ and $\{x^i\}$ are related to each other as follows:

$$t_{i_1' i_2' \cdots i_p'}{}^{j_1' j_2' \cdots j_q'} = L^{i_1}_{i_1'} L^{i_2}_{i_2'} \cdots L^{i_p}_{i_p'} L^{j_1'}_{j_1} L^{j_2'}_{j_2} \cdots L^{j_q'}_{j_q} t_{i_1 i_2 \cdots i_p}{}^{j_1 j_2 \cdots j_q}.$$

$$(A.1)$$

The valence—unqualified—of t is $p+q$. The customary term "rank" in place of "valence" is unfortunate. The matrix of components of a tensor of valence 2 already has a "rank." Note that provided $L^{i'}_i$ is taken to be $x^{i'}_{,i}$ here it does not matter whether it is constant or not.

11. Tensor densities. More generally, if t is a "tensor density of weight w and class k", it transforms according to the rule (A.1) nodified by the inclusion of an additional factor $|L|^{-w}(\text{sgn } L)^k$ on the right. Again L need not be constant.

12. Symmetries. A tensor, or tensor density, t is symmetric in r of its subscripts if the mutual interchange of any pair selected from these r indices leaves t unaffected. Symmetry or skew-symmetry in superscripts is similarly defined. If either $p=0$ or $q=0$ symmetry or skew-symmetry, unqualified, of t means that t is symmetric or skew-symmetric in all its indices.

13. Symmetrizing and alternating brackets. (i) Given t, select r of its subscripts or r of its superscripts and permute these in all possible ways. Add the $r!$ tensors so obtained and divide by $r!$. The result is a tensor (density) symmetric in the indices in question. It is represented by the same symbol as the original tensor (density), but with the indices in question enclosed in parentheses. For example, symmetrizing t_{ijkl} over j, k, l,

$$t_{i(jkl)} = \frac{1}{3!}\left(t_{ijkl} + t_{iljk} + t_{iklj} + t_{ijlk} + t_{ilkj} + t_{ikjl}\right). \tag{A.2}$$

Any index which is to be excluded is placed between bars. For example, symmetrization of t_{ijkl} over i, k, l, rather than j, k, l, leads to $t_{(i|j|kl)}$.

(ii) The process of skew-symmetrizing or alternating differs from that just described only in as far as each permutation must be multiplied by its sign before addition. The resulting object is then skew-symmetric in the indices in question which are now enclosed in brackets. For example, given t_{ijkl}, $t_{i[jkl]}$ is given by the right hand member of (A.2) provided the signs of the last three terms on the right are reversed.

14. The metric tensor and its signature. A Euclidian metric tensor has constant components, has covariant valence 2, is symmetric and is nondegenerate: $g_{[ij]} = 0$, $g := \det g_{ij} \neq 0$. By an allowed linear transformation $L^{i'}_i$ it can always be reduced to the form $g_{i'j'} = \text{diag}(e_1, e_2, e_3, e_4)$, where $e_i = 1$ or -1, as the case may be. Then $s := \Sigma^4_{i=1} e_i$ is the *signature* of g_{ij}. s is here required to have the value -2. The determinant g of the metric tensor is a tensor density of weight 2 and class 1, so that $\sqrt{|g|}$ has weight 1 and class 0. When the g_{ij} are not constant the definition of s remains valid locally.

15. Index juggling. A subscript (superscript) is raised (lowered) by transvection with the metric tensor. For example, if i is the index of interest and suppressing all others, then, given s_i, $s^i := g^{im} s_m$, or given t^i, $t_i := g_{im} t^m$.

Index juggling is the process of simultaneously raising one and lowering the other of a pair of dummy indices.

16. The permutation tensor and tensor density. Let ϵ^{ijkl} be a skew-symmetric tensor density with $w=1, k=1$. In a chosen coordinate system assign to ϵ^{1234} the value 1, which implies the values of all other components. Then these have the same values in any other coordinate system. ϵ^{ijkl} is called the contravariant permutation density. $e^{ijkl}:=|g|^{-\frac{1}{2}}\epsilon^{ijkl}$ often goes under the name "contravariant permutation tensor" which is a misnomer since although $w=0$ it has $k=1$ instead of $k=0$. Here, again, the constancy of g_{ij} is irrelevant.

INDEX